# SPENT

**Geoffrey Miller**

# S P E N T

Sex, Evolution, and Consumer Behavior

VIKING

VIKING
Published by the Penguin Group
Penguin Group (USA) Inc., 375 Hudson Street,
New York, New York 10014, U.S.A.
Penguin Group (Canada), 90 Eglinton Avenue East, Suite 700, Toronto,
Ontario, Canada M4P 2Y3 (a division of Pearson Penguin Canada Inc.)
Penguin Books Ltd, 80 Strand, London WC2R 0RL, England
Penguin Ireland, 25 St. Stephen's Green, Dublin 2, Ireland
(a division of Penguin Books Ltd)
Penguin Books Australia Ltd, 250 Camberwell Road, Camberwell,
Victoria 3124, Australia (a division of Pearson Australia Group Pty Ltd)
Penguin Books India Pvt Ltd, 11 Community Centre,
Panchsheel Park, New Delhi–110 017, India
Penguin Group (NZ), 67 Apollo Drive, Rosedale, North Shore 0632,
New Zealand (a division of Pearson New Zealand Ltd)
Penguin Books (South Africa) (Pty) Ltd, 24 Sturdee Avenue,
Rosebank, Johannesburg 2196, South Africa

Penguin Books Ltd, Registered Offices: 80 Strand, London WC2R 0RL, England

First published in 2009 by Viking Penguin, a member of Penguin Group (USA) Inc.

1 3 5 7 9 10 8 6 4 2

LIBRARY OF CONGRESS CATALOGING IN PUBLICATION DATA
Miller, Geoffrey.
Spent: sex, status, and the secrets of consumerism/Geoffrey Miller.
p. cm.
Includes bibliographical references and index.
ISBN 978-0-670-02062-1
1. Consumption (Economics) 2. Consumer behavior. I. Title.
HB801.M493 2009
339.4'7—dc22      2008051554

Printed in the United States of America
Set in Fairfield LH 45 Light
Designed by Francesca Belanger

*For my daughter, Atalanta Arden-Miller*

# Contents

# SPENT

# 1

## Darwin Goes to the Mall

CONSUMERIST CAPITALISM: it is what it is, and we shouldn't pretend otherwise.

But what is it, really? Consumerism is hard to describe when it's the ocean and we're the plankton.

Faced with the unfathomable, we could start by asking some fresh questions. Here's one: Why would the world's most intelligent primate buy a Hummer H1 Alpha sport-utility vehicle for $139,771? It is not a practical mode of transport. It seats only four, needs fifty-one feet in which to turn around, burns a gallon of gas every ten miles, dawdles from 0 to 60 mph in 13.5 seconds, and has poor reliability, according to *Consumer Reports*. Yet, some people have felt the need to buy it—as the Hummer ads say, "Need is a very subjective word."

Although common sense says we buy things because we think we'll enjoy owning and using them, research shows that the pleasures of acquisition are usually short-lived at best. So why do we keep ourselves on the consumerist treadmill—working, buying, aspiring?

Biology offers an answer. Humans evolved in small social groups in which image and status were all-important, not only for survival, but for attracting mates, impressing friends, and rearing children. Today we ornament ourselves with goods and services more to make an impression on other people's minds than to enjoy owning a chunk of matter—a fact that renders "materialism" a profoundly misleading term for much of consumption. Many products are signals first and material objects second. Our vast social-primate brains evolved to pursue one central social goal: to look good in the eyes of others. Buying impressive products in a money-based economy is just the most recent way to fulfill that goal.

Many bright thinkers have tried to understand modern consumerism by framing it in a historical context, asking, for example: How did we go from showing off our status with purple-bordered togas in ancient Rome to showing it off with Franck Muller watches in modern Manhattan? How did we go from the 1908 black Model-T Ford to the 2006 "Flame Red Pearl" Hummer? How did we go from eating canned tuna (about $4 per pound) to eating magical plankton ("marine phytoplankton, the ultimate nutrogenomic, supercharged with high-vibration crystal scalar energy healing frequencies"—$168 for fifty grams, or $1,525 per pound, from Ascendedhealth.com) as a luxury food?

This book takes a different approach from that of historical analysis. It frames consumerism in an evolutionary context, and thus addresses changes across much longer spans of time. How did we go from being small-brained semisocial primates 4 million years ago to being the big-brained hypersocial humans we are today? At the same time it addresses differences across species. Why do we pay so much for plankton, the most common form of biomass on the planet? Blue whales eat four tons of it per day, which would cost $12.2 million per day (plus shipping) from Ascendedhealth.com, if they wanted the "nutrogenomic supercharging."

To understand consumerist capitalism, it might help to begin by considering our lives today as our prehistoric ancestors might view them. What would they think of us? Compared with their easygoing clannish ways, our frenetic status seeking and product hunting would look bewildering indeed. Our society would seem noisy, perplexing, and maybe psychotic. To see just how psychotic, let's perform a thought experiment—something exotic, with time travel and lasers.

## From Cro-Magnons to Consumers

This is your mission, should you choose to accept it: Go back thirty thousand years in a time machine. Meet some clever Cro-Magnons in prehistoric France. (We'll assume that you'll be able to speak their language, somehow.) Explain our modern system of consumerist capitalism to them. Find out what they think of it. Would the prospect of ever-greater prosperity, leisure, and knowledge motivate them to invent agriculture, animal husbandry, walled towns, money, social classes, and conspicuous consumption? Or would they prefer to stagnate at their Aurignacian level of culture, knapping flint and painting caves?

Suppose you agree to this mission, and go back in your time machine. You find some Cro-Mags one evening, and get their attention by passing out a dozen laser pointers for them to play around with. After an hour they settle down, and you give your pitch, explaining that our culture offers a vast cornucopia of goods and services for showing off one's personal qualities in ten thousand new ways to millions of strangers. One acquires these displays of personal merit by "buying" them with "money" earned through "skilled labor." You promise that if they persist with their flint-knapping obsession, then in just a few millennia their descendants will be able to enjoy sophisticated cultural innovations, such as colonic irrigation and YouTube.

Your talk goes well, and it's time to gauge their reaction. You take some questions from the audience. One of the dominant adult males, Gérard, has been hooting with enthusiasm, and seems to get the idea. But Gérard has some concerns—most sound outrageously sexist to your modern ears, but since they are expressed with genuine curiosity, in the spirit of scientific objectivity you feel obliged to answer them honestly. Gérard inquires:

> So, Man-from-Future, with this money stuff, I could buy twenty bright young women willing to bear my children?

You: No, Gérard. Since the abolition of slavery, we can't offer genuine reproductive success in the form of fertile mates for sale. There are prostitutes, but they tend to use contraception.

Gérard: Well, I shall have to seduce the women so they want to breed with me. Can I buy more intelligence and charisma, better abilities to tell stories and jokes, more height and muscularity?

You: No, but you can buy self-help books that have some placebo effect, and some steroids that increase both muscle mass and irritability by 30 percent.

Gérard: OK, I will be patient and wait for my sexual rivals to die. Can I buy another hundred years of life?

You: No, but with amazing modern health care, your expected life span can increase from seventy years to seventy-eight years.

Gérard: These no-answers anger me, and I feel aggressive. Can I buy advanced weaponry to kill my rivals, especially that bastard Serge, and the men of other kin groups and clans, so I can steal their women?

You: Yes. One effective choice would be the Auto Assault-12 shotgun, which can fire five high-explosive fragmenting antipersonnel rounds per second. Oh—but I guess then the rivals and other kin groups and clans would probably buy them, too.

Gérard: So, we'd end up at just another level of clan-versus-clan détente. And there would be more lethal fights among hotheaded male teens within our clan. Then I shall be content with my current mate, Giselle—can I buy her undying devotion, and multiple orgasms so she never cheats on me?

You: Well, actually, lovers still cheat under capitalism; paternity uncertainty persists.

Gérard: What about Giselle's mother and sister—can I buy them kinder personalities, so they are less critical of my foibles?

You: Sadly, no.

Then Giselle, Gérard's savvy mate, interrupts with a few questions of her own, which you answer with ever-increasing dismay:

Giselle: Man-from-Future, can I buy a handsome, high-status, charming lover who will never ignore me, beat me, or leave me?

You: No, Giselle, but we can offer romance novels that describe fictional adventures with such lovers.

Giselle: Can I buy more sisters, who will care for my younger children as they would their own, when I am away gathering gooseberries?

You: No, child-care employees tend to be underpaid, overwhelmed, miseducated girls who care more about text messaging their friends than looking after the children of strangers.

Giselle: How about our teenage children—Justine and Phillipe? Can I buy their respect and obedience, and the taste to choose good mates?

You: No, marketers will brainwash them to ignore your social wisdom and to have sex with anyone wearing Hollister-branded clothing or drinking Mountain Dew AMP Energy Overdrive.

Giselle: *Zut, alors! Mange de la merde et meurs!* This money stuff sounds useless. Can I at least buy a mammoth carcass that never rots?

Finally, you see an opening, and you start explaining about Sub-Zero freezers—but then you remember that there is not yet an Electricité de France with fifty-nine nuclear reactors to supply freezer power, and you falter.

Giselle and Gérard are by now giving you looks of withering contempt. The rest of your audience is restless and skeptical; some even try to set you on fire with their laser pointers. You try to rekindle their interest by explaining all the camping conveniences that consumerism offers for the upwardly mobile Cro-Mag: sunglasses, steel knives, backpacks, and trail-running shoes that last several months, with cool swooshes on the sides.

The audience perks up a bit, and Giselle's mother, Juliette, asks, "So, what's the catch? What would we have to do to get these knives and shoes?" You explain, "All you have to do is sit in classrooms every day for sixteen years to learn counterintuitive skills, and then work and commute fifty hours a week for forty years in tedious jobs for amoral corporations, far away from relatives and friends, without any decent child care, sense of community, political empowerment, or contact with nature. Oh, and you'll have to take special medicines to avoid suicidal despair, and to avoid having more than two children. It's not so bad, really. The shoe swooshes are pretty cool." Juliette, the respected Cro-Magnon matriarch, looks you straight in the eye and asks, with infinite pity, "Are you out of your mind?"

## Contrasts and Choices

This thought experiment has, I hope, shaken your faith that humanity has ridden a one-way escalator of ever-increasing progress and ever-greater happiness since the Aurignacian. True, modern life can be a wondrous glee-glutted Funky Town for the wealthiest .01 percent of people on the planet. However, a fairer assessment would contrast the lifeways of an average prehistoric human and the lifestyle of an average modern human.

Consider the average Cro-Magnon of thirty thousand years ago. She is a healthy thirty-year-old mother of three, living in a close-knit clan of family and friends. She works only twenty hours a week gathering organic fruits and vegetables and flirting with guys who will give her free-range meat. She spends most of her day gossiping with friends, breast-feeding her newest baby, and watching her kids play with their cousins. Most evenings she enjoys storytelling, grooming, dancing, drumming, and singing with people she knows, likes, and trusts. Although she is only averagely intelligent, attractive, and interesting, most of her clan mates are too, so they get along just fine. Her boyfriend is also only average, but they often have great sex, since males have evolved wonderful new forms of foreplay: conversation, humor, creativity, and kindness. (About once a month, she hooks up secretly with her enigmatic lover, Serge, who has eleven confirmed Neanderthal kills, but whose touch is like warm rain on Alpine flowers.) Every morning she wakes gently to the sun rising over the six thousand acres of verdant French Riviera coast that her clan holds. It rejuvenates her. Since the mortality rate is very low after infancy, she can look forward to another forty years of life, during which she will grow ever more valued as a woman of wisdom and status.

Now consider the average American worker in the twenty-first century. She is a single thirty-year-old cashier, who drives a Ford Focus and lives in Rochester. She is averagely intelligent (IQ 100), having gotten Cs in a few classes before dropping out of the local community college. She now has this job in retail, working forty hours a week at the Piercing Pagoda in EastView Mall, fifty miles from her parents and siblings. She is just averagely attractive and interesting, so she has a few friends, but no steady boyfriend. She has to take Ortho Tri-Cyclen pills to avoid getting pregnant from her tipsy sexual encounters with strangers who rarely return her phone calls. Her emotional stability is only average, and because Rochester is dark all winter, she takes Prozac to avoid suicidal despair. Every evening she watches TV alone. Every night she fantasizes about being loved by Johnny Depp and being friends with Gwen Stefani. Every morning she awakens to the alarm clock next to the fake Chinese rubber plant in her six-hundred-square-foot apartment. It wears her out. Thanks to modern medicine, she can look forward to another forty-five years of life, during which she will become ever less valued as an obsolete health-care burden. At least she has an iPod.

By envisioning our current lives through our ancestors' eyes, we can see more clearly what we have given up, and what we have gained, from developing this thing called "civilization," which nowadays means consumerist capitalism. We can also better distinguish what is truly natural about our lives from what is historically accidental, culturally arbitrary, or politically oppressive.

Consumerist capitalism, as humans practice it in any particular culture, is not a natural or inevitable outcome of human evolution, given a certain level of technological sophistication. An evolutionary-psychology analysis of consumerism is accordingly not a way of giving science's seal of approval to consumerism, nor is it a way of morally justifying consumerism as the highest possible stage of biocultural progress. Many thinkers have tried to "naturalize" consumerism in that way, including most social Darwinists, Austrian School economists (Ludwig von Mises, Friedrich Hayek, Murray Rothbard), Chicago School economists (George Stigler, Milton Friedman, Gary Becker), Darwinian libertarians, globalization advocates, management gurus, and marketers. Their model (which I call the Wrong Conservative Model, because I think it's wrong, and because it's usually advocated by political conservatives) is:

human nature + free markets = consumerist capitalism

Against such attempts to "naturalize" consumerism, others have rejected any concept of "human nature" and any connection between biology and economics. These bio-skeptics include most Marxists, anarchists, hippies, utopians, New Age sentimentalists, gender feminists, cultural anthropologists, sociologists, postmodernists, and anti-globalization activists. For now, suffice it to say that such radicals propose the Wrong Radical Model, which is basically:

the blank slate + oppressive institutions + invidious ideologies = consumerist capitalism

Here, the "blank slate" means a human baby's big brain, allegedly born without any evolved instincts, preferences, or adaptations, yet capable of learning anything. (Steven Pinker trenchantly critiqued the possibility of such a brain in his book *The Blank Slate*.) The "oppressive institutions" are usually taken to be governments, corporations, schools, and media, as they inevitably represent the interests of some ruling class. The "invidious ideologies" are usually assumed to include religion, patriarchy, conformism, elitism, ethnocentrism, and mainstream economics. The Wrong Radical Model also usually assumes that Darwinism was invented as a justification for Victorian-era capitalism, including classism, colonialism, sexism, and racism—and if it's part of the problem, it can't be part of the solution.

As an alternative to the Wrong Conservative Model (consumerism as natural) and the Wrong Radical Model (consumerism as cultural oppression), this book proposes something a bit more complicated but, I hope, more accurate. I call it the Sensible Model, because I think it's pretty reasonable, given what science has discovered so far about people and societies. It goes like this:

> human instincts for trying unconsciously to display certain desirable personal traits
> + current social norms for displaying those mental traits through certain kinds of credentials, jobs, goods, and services
> + current technological abilities and constraints
> + certain social institutions and ideologies
> + historical accident and cultural inertia
> = early twenty-first-century consumerist capitalism

This more complex (but still vastly oversimplified) model does not just "denaturalize" consumerism. It also identifies specific things we could change about society by changing our social norms, institutions, ideologies, cultures, and technologies. The last third of this book suggests some possible ways to reengineer consumerist capitalism based on the Sensible Model.

These suggested changes will not aim to restore Cro-Magnon living conditions, which would be neither possible nor desirable for modern humans. There are 6.7 billion people on earth, and we can't all go back to living as hunter-gatherers. The notion of returning to an idealized paradise of simple, gentle, small-group living has been advocated by diverse visionaries throughout history: Buddha, Laozi, Epicurus, Thoreau, Engels, Gandhi, Margaret Mead, and the Unabomber. Often these visionaries attract followers, who form religions, political movements, or whole cultures: Taoists, Shakers, Luddites, Marxists, anarchists, hippies, and Emo kids. Even mainstream "bourgeois bohemians" support sustainability, voluntary simplicity, intentional living, organic farming, and corporate social responsibility, and try to smuggle some aspects of eco-communo-primitivism into their gated communities, insofar as local zoning permits them.

Yet each of these individuals and groups has exaggerated both the pros of primitive life and the cons of modern life. Each intuits correctly that a Cro-Magnon lifestyle was a more natural environment for the human body, mind, family, and clan. Yet at the same time, each forgets that, stripped of romantic idealization, Cro-Magnon life was also ignorant, insular, violent, and unimaginably boring. I would not want to live without civilization's key inventions—trade, currency, literacy, medicine, books, bicycles, films, duct tape, shipping containers, and computers. Unlike many malcontents, I consider the three best inventions of all time to be money, markets, and media. Each has radically increased the social and material benefits of peaceful human cooperation. But together they don't necessarily add up to consumerist capitalism in its current forms.

Fortunately, we are not forced to make an either-or choice between (1) eco-communo-primitivism as it might function in some elusive utopia, and (2) consumerist capitalism as it happens to have metastasized so far in some human societies. The Sensible Model suggests that there are many alternatives, and I think some of them combine the best natural features of prehistoric life and the best inventions of modern life.

## Mamas, Don't Let Your Babies
## Grow Up to Be Marketing Consultants

Cro-Magnons aside, modern society also looks bewildering to children. They are born with paleobrains, built from paleogenes, expecting a paleoworld: a close-knit social environment of kin-based hunter-gatherer clans. Children are wired to learn and play the normal game of life for which they evolved: be cute, grow up, find food, make friends, care for kin, avoid dangers, fight some enemies, find some mates, raise some kids, grow old and wise, die. Instead, they face a bizarre new world of frustrating duties and counterintuitive ideas: sit still, learn math, find a job, move away from friends, ignore kin, drive cars, leave kids in day care, and grow burdensome in old age.

Children face this new world with minimal guidance. Their parents go away all day to make money, to buy things, to look good and special, and to attract extra attention from other men and women, despite having mated and reproduced already. Their parents can't explain why they pretend that they're still in the mating market if they don't actually want a divorce and custody battle. Their high school teachers can't make sense of the consumerist world for them either, and their college professors can only suggest reading perplexing rants from postmodern French sociologists, such as Jean Baudrillard. So, almost everyone grows up confused, passes through life confused, and dies confused.

Only a few children do ever gain an intuitive grasp of consumerism's principles, and these typically grow up to be marketing consultants. They learn that people in general are motivated, at least unconsciously, to flaunt and fake their personal merits and virtues to one another. They realize that modern consumers in particular strive to be self-marketing minds, feeding one another hyperbole about how healthy, clever, and popular they are, through the goods and services they consume. Marketing consultants build careers around the postmodern insight: at its heart consumerist capitalism is not "materialistic," but "semiotic." It concerns mainly the psychological world of signs, symbols, images, and brands, not the physical world of tangible commodities. Marketers understand that they are selling the sizzle, not the steak, because a premium brand of sizzle yields a high margin of profit, whereas a steak is just a low-margin commodity that any butcher could sell.

However, even the cleverest marketers still don't fully understand which merits and virtues consumers are really trying to display through their consumption decisions. They don't really understand the content of the signals that people send to one another. Typically, marketers get some formal education in outdated consumer psychology research, then they get real jobs at real companies and realize that their formal training is mostly useless in selling real products. In response, they strive to develop an intuitive understanding of consumer behavior and marketing strategies through years of trial-and-error learning, plus the occasional book by Seth Godin or Malcolm Gladwell. They lack the huge practical benefits of having a coherent evidence-based theory about consumer behavior, and this limits their success rate.

In particular, most marketers still use simplistic models of human nature that remain uninformed by the past twenty years of research on human nature—research by evolutionary anthropologists, evolutionary biologists, and evolutionary psychologists. Marketers still believe that premium products are bought to display wealth, status, and taste, and they miss the deeper mental traits that people are actually wired to display—traits such as kindness, intelligence, and creativity. They don't put consumption in its evolutionary context, or trace its prehistoric roots, or understand its adaptive functions. As a result, they don't have access to a good map of the human mind, or of this brave new semiotic world in which it dwells. What marketers need is Darwin.

Yet Darwin, in turn, needs to take a break from fieldwork and visit the mall. The Darwinian science of human nature needs to shift some attention from Pleistocene evolution to twenty-first-century consumer behavior. We need to understand in much deeper ways how people flaunt and fake their biological fitness—their prospects for survival and reproduction—to one another. We need to understand the specific facets of fitness—the most important physical and psychological traits—that people strive to display through their "fitness indicators," including most of the products they buy.

## Fitness Indicators

Fitness indicators are signals of one individual's traits and qualities that are perceivable by other individuals. Almost every animal species has its own fitness indicators to attract mates, intimidate rivals, deter predators, and solicit help from parents and kin. Male guppies grow flaglike tails, male lions sport luxuriant manes, male nightingales learn songs, male bowerbirds build bowers, humans of both sexes acquire luxury goods. In each case, the fitness indicators are advertising fundamental biological traits such as good genes, good health, and good social intelligence.

The animals that possess them are not consciously aware that these traits evolved to advertise their fitness. They just have the genes and instincts for displaying them, and evolution itself keeps track of the survival, social, and sexual benefits of doing so. We humans may not have much more conscious insight into the biological functions of our fitness indicators than guppies have into the functions of their flaglike tails. Indeed, we often buy products that increase our apparent fitness (health, beauty, fertility, intelligence) at the cost of real biological fitness (reproduction)—for example, Ortho Tri-Cyclen birth control pills make women's skin look more attractive by reducing acne, but it lowers reproductive success by eliminating ovulation. Our brains did not evolve to pursue reproductive success consciously, but to pursue the cues, experiences, people, and things that typically led to reproductive success under ancestral conditions.

Successful reproduction requires males and females to follow different sexual strategies, and to display their fitness indicators to different audiences. Across virtually all animal species, males display mostly to attract female mates, and less often to intimidate male sexual rivals. It is easy to see the functional connection between peacock tails and Porsches, and many recent studies have confirmed that men increase the conspicuousness of their consumption when they are most interested in mating. The situation is more complex for females. Female animals of most species gain little benefit from displaying fitness indicators to either sex, except in species where females compete for resources and mates, or where males are selective about their mates. Among the highly social great apes, for example, female status hierarchies are important in predicting female access to food, so female apes often compete for status by displaying fitness indicators to one another—such as their size, health, assertiveness, and popularity during mutual grooming. Such female-versus-female status competition probably likewise accounts for most conspicuous consumption by human females, especially for products such as Prada handbags and Manolo Blahnik shoes, which straight males rarely notice. Humans are even more distinctive in that males are fairly choosy about the females with whom they form long-term relationships, which means that females also compete to attract the higher-quality males. Sadly, the evidence so far suggests that men pay very little attention to such conspicuous consumption by women.

Unlike other animals, humans have evolved unique abilities to invent, make, display, and imitate new kinds of fitness indicators. These new indicators evolve at the cultural rather than genetic level, and they include many of the credentials, jobs, goods, and services that are typical in modern economies. Juvenile humans have an insatiable thirst to learn about culture-specific indicators, gossiping endlessly about what is "cool," "hot," "phat," "rad," or "wicked." In other words, they are trying to discern "Which products would display my traits, tastes, and skills most effectively, given the current display tactics favored by my peer group, especially its more socially and sexually attractive members?" If local status depends on memorizing longer passages of the Torah or Qur'an than others can, young people will learn to do that; if it depends on getting higher "interestingness" scores on one's Flickr-posted photos, or a higher friend-count on Facebook, or higher "hotness" ratings on Hotornot.com, they will opt for those instead. Just as toddlers have special brain systems that evolved to learn whatever language is spoken locally, teens seem to have evolved similar systems to learn whatever culture-specific fitness indicators are favored in their local eco-niche, social niche, or market niche. We are not just intuitive linguists; we are also intuitive status-ticians. In each case, evolution has crafted our innate ability to acquire culturally modulated communication skills.

In humans, fitness indicators are unlikely to have evolved to advertise monetary wealth, career-based status, or avant-garde taste, because these phenomena arose quite recently on the evolutionary timescale, within the past ten thousand years. Rather, the key traits that we strive to display are the stable traits that differ most between individuals and that most strongly predict our social abilities and preferences. These include physical traits, such as health, fertility, and beauty; personality traits, such as conscientiousness, agreeableness, and openness to novelty; and cognitive traits, such as general intelligence. These are the biological virtues that people try to broadcast, with the unconscious function of attracting respect, love, and support from friends, mates, and allies. Displaying such traits is the key "latent motive" that marketers strive to comprehend. While consumers do strive semiconsciously to show off their wealth, status, and taste, I'll argue that they do so largely in order to reveal these more fundamental biological virtues. Certainly, money can function as a form of "liquid fitness," but largely as a means of acquiring more conspicuous fitness indicators. And while consumers do rely more on emotions than on reason in deciding what to buy, human emotions cannot be described clearly without understanding their evolutionary origins and functions. Until marketers and consumers understand these principles deeply, in vivid Technicolor detail, and with a bittersweet ambivalence about the human condition, we will have little hope of improving and enlightening society.

## Description and Prescription

This book has two main aims. The first is to describe our human culture as it is, within a biological context. The second is to suggest some ways that we could change our human culture so it more happily combines the best features of prehistoric social life and modern technology.

Inevitably, my descriptions and prescriptions will get mixed together in the course of my discussion. They will interlace at every scale, from recurring book themes down to specific product examples, as I shift between considering facts and values. Such promiscuous hybridizations of "is" and "ought" often provoke outrage among the superhumanly rational philosophers of science and morality, who prefer that behavioral scientists restrict themselves to objective reportage and leave the preaching to them, or to their religious counterparts. Too bad. There is a distinguished tradition of gaining new prescriptive insights into one's society through new ways of describing its follies and injustices—a tradition that includes such names as John Locke, Mary Wollstonecraft, Daniel Defoe, William Wilberforce, Henry David Thoreau, Karl Marx, Max Weber, Margaret Sanger, Thorstein Veblen, John Kenneth Galbraith, Alfred Kinsey, Germaine Greer, and Peter Singer. I hope to scramble along like a dormouse in their footsteps.

If my descriptive analysis proves accurate, it should be useful to different readerships with conflicting agendas. It should give marketers new ways to exploit consumer preferences and make more money. It should also give consumers new ways to resist marketer influence and save money. It may give conservatives new ways to justify some aspects of the status quo, given the ubiquity of conspicuous display throughout nature. It may also give progressives new ways to undermine that status quo, given the colossal inefficiency of conspicuous consumption as a form of trait display. While I can't control who reads this book, what insights they derive from it, or how they apply those insights in their lives and livelihoods, I can hope that a more accurate view of human nature and consumerist culture leads to more intelligent debate about all its relevant issues.

## Consumerist Ambivalence

Like most reasonable people, I feel deep ambivalence about marketing and consumerism. Their power is awe-inspiring. Like gods, they inspire both worshipful submission and mortal terror. Consumerist capitalism produces almost everything that is distinctively exciting about modern life and almost everything that is appalling about it. Most people like clothing, shelter, safety, education, medicine, and travel, and would miss them if they lived in an eco-communo-primitivist utopia. Most people dislike exploitation, workaholism, runaway debt, pollution, the military-industrial complex, cartels, corruption, alienation, and mass depression, and would not miss them. Then there are personal tastes. The things I find most exciting about consumerist capitalism include: almond croissants, Tori Amos concerts, skiing at Telluride, houses designed by Bart Prince, the BMW 550i, Provigil, iPods full of Outkast and Radiohead songs, and the Microsoft Ergonomic keyboard on which I'm typing. The things I find most appalling: Las Vegas, the Mall of America, fast food, cable television, Hummers, and overpriced phytoplankton. Then there are the things that seem both exciting and appalling: frappuccinos, business schools, *In Style* magazine, Glock handguns, Jerry Bruckheimer movies, Dubai airport duty-free shops, Diet Code Red Mountain Dew, the contemporary art market, and Bangkok. You can draw up your own lists, and contemplate your own sources of consumerist ambivalence.

Unfortunately, most writing about consumerism shows either pure love or pure hate, with no balance or nuance. On the one hand, we have pro-consumerism advocacy: the World Trade Organization, World Bank, and World Economic Forum; the *Economist* and the *Wall Street Journal;* marketers, corporate lobbyists, and libertarians. On the other hand, we have anticonsumerism activism: Greenpeace, Earth First, Naomi Klein's *No Logo, Adbusters* magazine, the New Urbanism, voluntary simplicity, the Slow Food movement, the Fair Trade movement, Buy Nothing Day, and True Cost Economics.

The extremism in either case is . . . extreme. Both sides have been shouting past each other for decades. My goal here is not to conduct a cost-benefit analysis of consumerism, or to reach some simplistic good versus bad judgment. Rather, my hope is that by grounding our understanding of consumerism in the biological realities of human nature and individual differences, pro-consumerism and anti-consumerism advocates can find a higher, closer, common ground. It's not enough to recognize that both sides have some good points and good intentions. We need to step back from the contemporary debate and reassess it from the broadest, deepest possible perspective—not only from a cross-cultural, historical perspective, but also from a cross-species, evolutionary perspective.

## 2

# The Genius of Marketing

THE TALE OF ALADDIN, from the *1,001 Arabian Nights*, offers a serviceable metaphor for consumerism. The poor boy Aladdin discovers a magic lamp in a secret cave. When he rubs the lamp, he releases a terrifying but powerful Genius ("genie," colloquially). This Genius of the Lamp grants Aladdin many wishes—rich food on silver plates, embroidered clothes, fine horses, the intimidation of a sexual rival (the Grand Vizier's son), forty gold basins, and a marble palace set with jasper, gold, and rubies. In Aladdin's case, the reproductive payoffs of releasing and mastering the Genius are real: he wins the love of a princess and sires a long line of kings.

In the modern world, the market is the Genius, and its products embody our wishes, though the biological payoffs in this case are less clear. The market holds a mirror up to our desires, creating public manifestations of our private preferences. Like the Genius of the Lamp, the market seems both magical and manic in its potency and ingenuity. Through many cycles of market research, consumer feedback, and economic competition, the market, like the Genius, also makes enormous efforts to fulfill our stated wishes—but often, like the Genius, it obeys the letter rather than the spirit of those wishes, with frustrating consequences.

Marketing has already transformed our world more dramatically than the Genius transformed Aladdin's. Thirty thousand years ago, we could learn very little about ourselves by looking outside at our environment. There were rocks, trees, insects, and stars—an ornery reality from which we had to wrest a living. Yet in the twenty-first century, at least for educated elites in rich countries, consumer capitalism has profoundly reshaped our environment to reflect our wishes. So, to understand the nature of our wishes, we need only look out at the world and see what it says about us. The world of goods, services, advertisements, media, and entertainment is a rich source of evidence about what people want—or at least about the products that people think they want.

Roughly, products fall into two overlapping categories: (1) things that display our desirable traits and bring us "status" when others see that we own them, and (2) things that push our pleasure buttons and bring us satisfaction even if no one else knows we have them. This book focuses on the first category, the status products, which reveal human instincts for displaying various traits to others. Analyzing such products can even help us discern the nature of the human traits they are designed to display. For instance, to better understand the trait of human intelligence, we can go beyond traditional IQ tests, and analyze ways that bright consumers acquire educational credentials (Oxford MAs, Harvard MBAs) as intelligence displays. To better understand the trait of human altruism, we can go beyond experimental economic studies of the well-known Prisoner's Dilemma and Ultimatum Game, and analyze the ways some consumers unconsciously display their kindness by driving their Toyota Camry hybrids to their local organic food co-ops to pick up their Fair Trade shade-grown coffee. For every aspect of human nature, for every dimension of variation in human personality, intelligence, virtues, and values, there exists a vast market of product sets that we can draw from to broadcast our personal traits to others.

The same reasoning applies to pleasure products. To understand our aesthetic tastes, we can go beyond laboratory research on visual preferences, and examine the clothes and cars designed to attract the human eye. To understand male sexual psychology, we can do more than ask men what they're willing to say they want; we can also analyze the ways that female sex workers have learned to maximize their earnings by looking and acting in certain ways. Many evolutionary psychologists have already been working for years to better understand human tastes and preferences by analyzing such pleasure products; the status products remain much more mysterious.

To understand how human status seeking plays out in consumerist capitalism, we need a new way of thinking about the human condition, one that goes beyond conventional science and conventional ideas about consumerism. Most writing about consumerism assumes that culture shapes human nature, so that our desires conform to the dictates of advertising, through socialization and learning. This is the heart of postmodernist cultural theory. Much of evolutionary psychology tends to work in the same direction, from the outside in, as Darwinians consider how the external challenges of prehistoric living would have shaped our thoughts and feelings through genetic evolution. Both postmodernists and Darwinians agree that our minds conform to the external environment, through either cultural or biological adaptation. I try to work in the complementary direction, from the inside out. I argue that we have inherited a rich human nature from our ancestors, full of desires and preferences for seeking status and impressing others. I recognize that we have a few key dimensions of variation in intelligence and personality that are genetically inherited, sexually attractive, and socially valued, and that these drive most consumerist displays. I try to trace how our internal status-seeking instincts get refracted through consumerist culture to produce the products, markets, and lifestyles that constitute our modern environment.

A deeper understanding of human nature helps everybody, whether we are consumers trying to live a more fulfilling life, or marketers trying to increase brand recognition and market share, or scientists trying to understand the world, or activists trying to improve society. Indeed, new insights into human nature have provoked every other major revolution in the classical liberal tradition of evidence-based enlightenment: the Protestant Reformation, the American Revolution, the abolition of slavery, and the women's movement. Those democratic revolutions reformulated the relationships between governors and citizens, but with the spread of free-market capitalism, business-consumer relation-ships have become much more important. What democracy is to poli-tics, consumer demand is to business: the fulcrum with which ordinary people have the most leverage on how their world is organized. *Spent* is therefore not just a book about business or psychology; it concerns the most important issue that confronts a free-market society—how to make the economy work for us, rather than vice versa.

## Beyond Maslow

If we want to understand our behavior as consumers, it helps to remember that we evolved as social primates competing for mates, friends, family support, and status. Throughout most of the twentieth century, psychologists assumed that this biological legacy gave us just a few simple instincts to survive and reproduce, and that everything else was due to learning and culture. Sigmund Freud, Jean Piaget, and B. F. Skinner had some great insights, but they did not integrate Darwin's legacy into human psychology. This Darwinization has taken place only since about 1990, in the field of evolutionary psychology, and it has yielded a picture of human nature much richer than a short list of simple instincts.

Evolutionary psychology reaches all the way into our most cherished abilities and aspirations, explaining why we care about friendship, love, family, social status, self-respect, moral virtue, and authenticity. It explains much more than merely why "sex sells." It also explains why empathy and extraversion are sexually attractive, and why we are moti-vated to buy mobile phones that reveal our popularity, and pets that reveal our kindness and conscientiousness. It has demonstrated a set of human motivations and aspirations vastly more detailed, nuanced, and principled than the "hierarchy of needs" posited by Abraham Maslow in the 1950s, which remains the dominant model of human motivation in consumer-behavior textbooks.

Maslow's hierarchy includes just seven types of human needs, clustered into two categories. "Deficiency needs" are drives to reduce states of deficiency or discomfort, and are pursued only when a deficiency arises. They include:

- physiological needs: breathing, drinking, eating, excreting, regulating temperature, having sex
- safety needs: health, well-being, familiarity, predictability, personal security, financial security, insurance
- social needs: family, friendship, intimacy, sexual love, belonging, acceptance
- esteem needs: recognition, status, fame, glory, self-respect, self-esteem

"Growth needs" are drives toward "transcendence," and are pursued whenever the individual is free to do so. They include:

- cognitive needs: to learn, explore, discover, create, acquire knowledge, and increase intelligence
- aesthetic needs: to experience beauty as found in nature, people, or artifacts
- self-actualization needs: to fulfill one's potential and make the most of one's abilities

From an evolutionary viewpoint, Maslow's hierarchy is hopelessly muddled. It mixes innate drives (breathing, eating, seeking status, acquiring knowledge) and learned concerns (seeking financial security, self-esteem, and increased intelligence). It does not "cut nature at the joints" in terms of the key selection pressures that shaped human behavior: survival and reproduction. Survival includes most of Maslow's physiological needs (breathing, eating), but also some of the more concrete safety needs (avoiding harm from predators, parasites, sexual rivals, and hostile tribes), social needs (building relations with family, friends, and mates who can help feed, protect, and heal you under adverse conditions), cognitive needs (to learn about survival-increasing opportunities and survival-reducing dangers), and even aesthetic needs (to find a propitious landscape for one's clan to live in, to make weapons that are serviceably symmetric, strong, and sharp). Reproductive challenges, including finding high-quality sexual partners and raising high-quality offspring, encompass one of the key physiological needs (having sex) and most of the other social, esteem, cognitive, aesthetic, and self-actualization needs. For example, mate preferences for kindness can explain our social needs for intimacy, belonging, and acceptance. Mate preferences for status can explain our esteem needs for recognition, fame, and glory. Mate preferences for intelligence, knowledge, skills, and moral virtues can explain our cognitive needs to learn, discover, and create, and our self-actualization needs to fulfill our potential (for example, to display the highest possible mate value given our genetic quality).

Moreover, a branch of evolutionary theory called "life history theory" points out that there are often tough trade-offs between these survival and reproductive priorities. The lower-level needs do not always take priority. For example, male elephant seals will often starve to death during a breeding season while guarding their harems. If elephant seals could talk, and you recruited them to participate in a focus group at your market research institute, they might explain that they were giving up a physiological need (to eat) for three higher needs: a social need (to feel intimacy and belonging with each of many females), an aesthetic need (to be surrounded by beautiful—that is, fine, fit, fat, fertile—females), and a self-actualization need (to be the best elephant seal one can be, as demonstrated through biting, mauling, bloodying, and excluding all male sexual rivals from one's beachfront harem). But these last three Maslovian needs can actually be reduced to reproductive benefits. Natural selection crafted social, aesthetic, and self-actualization motivations because they yielded higher reproductive success over thousands of generations of elephant seal evolution. Male elephant seals who were "slackers," content to fulfill their survival and safety needs without conflict, would have avoided the bloody beach sites where more ambitious "status seekers" fought, copulated, starved, and died. The slacker seals may have been perfectly happy, and might have even turned vegan and ate plankton, but they did not leave any descendants to inherit their easygoing temperaments. Only the male seals that were willing to compete for dominance, status, and harems, even at the cost of their own lives, sired any offspring. Although male humans did not evolve in prehistory to compete for large harems, both human sexes did evolve to compete for high-quality mates, friends, and allies, leaving us with many of the same drives, instincts, preferences, and aspirations that the elephant seals in the focus group might have articulated.

Finally, Maslow's hierarchy overlooks most of the adaptive preferences, emotions, motivations, and aspirations that evolutionary psychology has demonstrated in human nature. It conflates different forms of love—parental solicitude toward offspring, familial solicitude toward kin, social attachment to same-sex friends, romantic attachment to mates, cultural attachment to one's tribe. It ignores the distinctive functions of gratitude, guilt, shame, embarrassment, moral outrage, and forgiveness in sustaining cooperation within groups.

While Maslow's work was a useful early step in categorizing the diversity of human motivations, it never integrated Darwinian insights, and it is now seriously outdated. Its continuing popularity in marketing and consumer behavior textbooks is puzzling, especially since marketing professionals actually don't use it much in their day-to-day thinking about consumer behavior; perhaps nothing better has yet emerged to take its place. As our understanding of human nature has become broader, deeper, and more subtle over recent years, we should have been able to understand an ever-greater range of products, marketing issues, and consumer behavior patterns. Why hasn't this happened?

## Why Evolutionary Consumer Psychology Is Just Getting Started Now

Over the past two decades, evolutionary psychology has been offering new insights into our motivations, emotions, preferences, relationships, and even aesthetic tastes. College students have learned about this new science through the hundreds of evolutionary psychology courses that are springing up across North America, Europe, and Asia. The public has become familiar with it through excellent popular books by Richard Dawkins, Steven Pinker, David Buss, Matt Ridley, E. O. Wilson, and others, and through fascinating TV documentaries by PBS and the Discovery Channel in the United States, and by the BBC and Channel 4 in the UK.

Evolutionary principles have also revolutionized many traditional disciplines during the same period. Hundreds of papers and dozens of books have discussed Darwinian medicine, Darwinian psychiatry, evolutionary analysis in law, evolutionary economics, Darwinian political science, Darwinian aesthetics, and Darwinian moral theory. Insofar as evolutionary principles promote a more coherent understanding of human nature, their popularity should not be surprising, because human nature is at the foundation of all social sciences and humanities. However, the business world has conspicuously remained the odd man out. Marketing, advertising, consumer research, and product development are equally dependent on an accurate understanding of human nature, yet evolutionary insights have so far had little influence in this realm.

Executives are still trained in MBA programs, and market researchers are still trained in Ph.D. programs, as if humans were created from clay eight thousand years ago, and designed with an arbitrary list of "manifest motives" and "latent motives." Virtually no course content on the evolutionary origins of human behavior and preferences is included at any of the world's top business schools—IMD (Lausanne), INSEAD (Paris), ESADE (Barcelona), London Business School, Rotterdam School of Management, Indian Institute of Management (Bangalore), Queens School of Business (Toronto), Harvard, Stanford, MIT (Sloan), U. Penn (Wharton), New York University (Stern), or Northwestern (Kellogg). To date, only a few researchers have used Darwinian insights in any systematic way to understand consumer behavior.

Since the late 1990s the marketing professor Gad Saad, at Concordia Business School in Montreal, has been developing this new field of evolutionary consumer psychology almost single-handedly. He published the first papers about evolutionary psychology to appear in any marketing or consumer behavior journals, and published the first book on the topic, *The Evolutionary Bases of Consumption*, in 2007.

Since the mid-1980s, the Cornell economist Robert H. Frank has been using evolutionary principles of social and sexual competition to understand the more specific issues of runaway economic status seeking and conspicuous consumption. His books, such as *Choosing the Right Pond*, *The Winner-Take-All Society*, and *Luxury Fever*, have not only connected Darwin and Veblen, and framed human economic behavior in its biological context, but have also pioneered new empirical ways to analyze economic data, in order to demonstrate the pervasive effects of status seeking in career choices and consumer choices. (Robert H. Frank is not to be confused with journalist Robert L. Frank, author of *Richistan*.) I owe Gad Saad and Robert Frank a great debt for their groundbreaking work.

More recently, a few other researchers, such as the marketing professors Vladas Griskevicius at University of Minnesota and Jill Sundie at University of Houston, have been taking evolutionary consumer psychology forward in new directions by integrating it more closely with social psychology. A few other evolutionary psychologists have thought about human nature in relation to particular kinds of products, such as food, pets, landscapes, singles ads, drugs, pornography, and novels. In each case, by more clearly understanding the evolutionary origins, biological functions, and design features of our psychological adaptations (such as our perceptions, emotions, and preferences), researchers can better understand the "hedonomics"—the pleasure-giving design features—of various goods and services.

At the moment, though, Darwinians have only scratched the surface of consumer behavior. The most powerful theory in the whole of the biological and behavioral sciences, a theory that explains the origins and functions of the complex psychological adaptations that constitute human nature, has rarely been called on to illuminate the swamps and jungles of modern consumerism, where we all live these days. For example, given the antibiological biases of most consumer-behavior researchers and journal editors, as of mid-2008 only one paper mentioning evolutionary psychology has ever appeared in any of marketing's four leading academic journals—*Journal of Consumer Research, Journal of Marketing, Journal of Marketing Research*, and *Marketing Science*. None of them has ever published an article concerning biological evolution, human nature, Darwinism, or primate behavior.

Consumer research has been almost as oblivious to the extraordinary recent progress in the study of individual differences—the ways that people's minds are distinct from one another's. Individual-differences research has delivered some wonderfully robust and useful models of human personality, intelligence, and moral virtue. These models are much simpler than might have been expected. Human personality, for example, can be represented quite accurately by just the "Big Five" dimensions of variation between people: openness to experience, conscientiousness, extraversion, agreeableness, and emotional stability. Human intelligence can be represented with astonishing efficiency and accuracy by just one dimension, called the $g$ factor (a.k.a. general intelligence, general cognitive ability, IQ). As we'll see later, if we know how an individual scores on these "Central Six" dimensions (the Big Five personality traits plus general intelligence), we can predict a great deal about his habits, preferences, values, and attitudes—and about the products he may acquire to display those traits to others. All six dimensions are also genetically heritable: twin and adoption studies show that these individual differences are predicted at least moderately by genetic differences, and not just by family upbringing or random effects during development. They are all fairly stable across the life course, so one's score in adolescence reasonably predicts one's score in older age. They are all salient to other people during normal social interaction, and are assessed fairly accurately, if unconsciously, even within the first few minutes of interaction with a stranger. While some recent textbooks on consumer behavior and marketing have started to pay lip service to the Big Five traits in a paragraph or two, these traits are still almost never mentioned in popular marketing books, or used in marketing practice. Discussions of general intelligence remain taboo throughout marketing theory and practice.

The advances in evolutionary psychology and individual differences research have rarely been used to understand consumerism, because very few consumer researchers understand the new psychology, and very few psychologists know anything about marketing, advertising, or product development. It is admittedly difficult to straddle the worlds of science and business. Science strives for cumulative progress through humbly authoritarian respect for one's predecessors (through citations) and colleagues (through collaboration and peer review), whereas most new business books pretend to offer 100 percent fresh, new, radical, and unprecedented concepts, enabling its authors to profit from corporate speeches and consulting work. Science tries to build coherent, nuanced, testable theories that look gravely intimidating, whereas business books offer bullet-point lists and two-by-two graphs that look winsomely simple. Scientists try to use consistent technical terms that can be understood by them and no one else, whereas business books invent wacky new catchphrases that sound great but that can't really be understood by anyone (*Gung ho! The millionaire mind! Who moved my cheese? Lead like Jesus! Eat that frog! Purple cow!*). If you're accustomed to reading popular business books that mimic the ADHD pace of Jerry Bruckheimer action films, you'll need to adopt a calmer reading mode here—one that allows, I hope, some elbow room for thinking, judging, and reflecting. On the other hand, if you're used to reading scientific journal papers, you'll just have to hang on as my writing careens pinball-style from topic to topic, and seek your quietude in the copious notes and references on the book's website.

## This Book

*Spent* concerns where we are today—how we live within this wondrous, horrific, perplexing world of consumerist capitalism that we have built over the past few generations—and where we could go in the future. My first book, *The Mating Mind*, concerned where we come from—how our ancestors lived in prehistory and how human nature evolved over the past few million years. It argued that some of our most wondrous and distinctive human mental abilities—art, music, language, kindness, intelligence, and creativity—evolved not just for survival, but for reproduction. Specifically, they evolved in both sexes as fitness indicators to attract high-quality sexual partners.

To explain how the process of sexual selection through mate choice might have shaped human mental evolution, *The Mating Mind* used a lot of marketing metaphors. Animals seek sexual partners in a competitive mating market. Animal bodies and behaviors evolve largely as advertisements for their genes. Male humans evolved potent new sales tactics—verbal courtship, rhythmic music, gentle foreplay, prolonged copulation—for seducing skeptical female customers into accepting free trials of their fastest-moving consumer goods (sperm). Female humans evolved potent new tactics of relationship marketing to build long-term loyalty among their highest-value male customers, and to promote continued male investment in their new subsidiaries (children). Human creativity evolved to keep our mates fascinated as we release ever-new behavioral products—new utterances, stories, jokes, observations, ideas, artifacts, songs, and gifts—designed to seem initially fashionable but that quickly become obsolete. Each individual's ideology (religious, political, and philosophical beliefs) can even be viewed not as his editorial content but as his ad campaign—designed not to convey verifiable news about the world, but to create positive emotional associations between the individual as product and the customer's aesthetic, social, and moral aspirations.

These marketing metaphors seem to work as well as they do because most readers know more about shopping than about sexual selection theory, so the latter could be explained by reference to the former. *Spent* tries to reverse the direction of explanation, by analyzing consumer behavior based on what we know about human evolution and individual differences. This task might prove harder, given that it entails explaining the apparently familiar in terms of the unfamiliar, as if one were to say, "Look, it's really very simple to draw a dog; you simply visualize the molecular structure of ethanol, and imagine the oxygen atom is the dog's cranium, and the two carbon atoms form the dog's torso . . ." Nonetheless, it's worth a try, because we really need to understand how consumerist capitalism arose from human nature, and how it could be improved.

To follow my reasoning, you'll need to rethink most of what you thought you knew about your motives, preferences, and aspirations. You'll have to look at your adult human life the way a wise child or a Cro-Magnon matriarch would. You'll have to set aside some traditional distinctions between biology and culture, animals and consumers, evolution and economics, psychology and marketing. You'll need some existential courage to accept that years of obsessive workaholism and status-seeking consumption may have been misguided.

That's the hard part. The easy part is that *Spent* demands very little background expertise. You don't need to know much about psychology, beyond what you already know about people. You don't need to know much about consumerist capitalism, beyond what you already know about shopping. In fact, the less you've been taught about traditional marketing and economics, the fewer misconceptions you'll have to overcome.

My ideas will also be easier to follow if you haven't been taught too much cultural theory, postmodern philosophy, gender feminism, cultural anthropology, media studies, or sociology. While these fields have produced most of the trenchant thinking and writing about consumerism, they usually preach that human biology has nothing to do with human culture, consumption, or ideology. They usually preach that scientists work to maintain the status quo, and that evolutionary psychologists like me are especially dangerous and conservative. Even many marketers have been socialized to take that view. As you will see, such preaching is false. Evolutionary psychology can offer a deeper, more radical critique of consumerist culture than anything developed by Marx, Nietzsche, Veblen, Adorno, Marcuse, or Baudrillard. We can respect their insights without insisting that they're more profound than Darwin. We can combine their moral outrage, playful irreverence, and utopian imagination with the best of twenty-first-century science, and see how far we can get.

At the practical level, I'll consider mostly goods and services from companies with recognizable brands, websites, and advertisements that readers can buy retail for a fairly standard cost, that interest people across a wide range of sexes, ages, cultures, and countries, and that can be illuminated by evolutionary psychology and individual differences research. Most specifications and prices for particular products are from the company websites or print advertising as of 2007 or 2008. I devote less attention to many product categories that are economically important but less interesting, such as commodities and raw materials (steel, oil, plastic, lumber, grain), basic domestic utilities (water, gas, electricity, heating, cooling, lighting), basic consumer durables (appliances, furniture, linens), and financial products (banking, credit, mortgages, insurance, equities, bonds, wills, trusts). In many of these categories, consumerist showing off and status display are less important, as when one is seeking the best price on soybean oil futures, or the best heart surgeon, or the most reliable life insurance company. No doubt the evolutionary psychology of consumer behavior will eventually embrace all these product categories, but I won't here.

## This Author

Cultural theorists have proposed a good insight: books are easier to understand when their authors are candid about their background and motives, and self-critical about their likely biases and blind spots. Because evolutionary psychologists have often been caricatured as racist, sexist, conservative reductionists, it's especially important to clear away those misconceptions. Just for the record, I'm a secular humanist, an antiwar internationalist, an animal-rights environmentalist, a pro-gay feminist, a libertarian on most social, sexual, and cultural issues, and a registered Democrat—in other words, a typical psychology professor.

At the University of New Mexico, I work with about half a dozen

Ph.D. students, doing research on human mate choice, intelligence, creativity, personality, mental illness, humor, and emotions. My wife and I have a twelve-year-old daughter, a thirteen-year-old Toyota, and a fifty-four-year-old house in Albuquerque. I try to understand the crushing poverty and impotent despair that still afflicts half of our species (including most people in South America, Africa, Asia, and graduate school), but my modest tenured income means these problems aren't very salient to me. I was born two-thirds of the way through the twentieth century, so I'm too old to care about cell-phone fashions, and too young to care about hospice costs. Like 1.47 percent of humans on earth, I'm a white heterosexual American male. So, I try to be a good Darwinian feminist, but my sex and sexual orientation mean I'll slip sometimes. I've lived abroad for nine years, but only in England and Germany. I try to be globally aware, but my race, nationality, and limited expat experience mean I tend to overlook many issues.

Culturally, I'm eclectic and ambivalent. I enjoy anticonsumerism books by Thomas Frank and Juliet Schor, but I also subscribe to the *Economist* and *Wired*. I enjoy lefty-radical-feminist music by Ani DiFranco and Tori Amos, but I have immense respect for the business world, and gratitude to the workers, managers, and investors who provide our necessities, luxuries, and entertainments. I appreciate that the Prius exists, but I drive a tanklike Land Cruiser (and you would, too, if you saw how people drive in Albuquerque). I loathe malls, but I respect the free market as the most ingenious system yet devised for people to enjoy mutual gains from trade under conditions of peace, freedom, and autonomy. I hate the way that corporate lobbyists corrupt democracy, but I recognize that our quality of life in the developed world is a fragile, fortunate exception to the global historical norm of toil, oppression, poverty, disease, and death.

My interest in this topic came from two intellectual awakenings—one around 1990 concerning evolutionary psychology's power to explain human nature, and one around 2000 concerning marketing's power in modern culture. In 1988 I was a psychology Ph.D. student at Stanford University, after a B.A. at Columbia University in New York and a childhood in Cincinnati, Ohio. That year, some key founders of evolutionary psychology—Leda Cosmides, John Tooby, David Buss, Martin Daly, and Margo Wilson—were all visiting Stanford for a sabbatical year. My friend Peter Todd and I got excited about their ideas, met with them about once a week, and learned about the extraordinary potential of Darwinian theory to revolutionize psychology. Everything about human behavior suddenly seemed easier to understand—clearer, simpler, more functional, more grounded in the 3-billion-year saga of life on earth. Everything in psychology seemed more unified—more connected not only to the other sciences, but also to the humanities and to everyday life. I got hooked on the idea that human behavior could be best understood by considering the challenges of survival and reproduction that our prehistoric ancestors faced. This paradigm shift seemed uniquely satisfying and complete—as if I had found my intellectual home once and for all, and nothing could ever blow my mind in the same way again.

Fortunately, I was wrong. About ten years later, I got a research job at the new Centre for Economic Learning and Social Evolution at University College London. My challenge was to get the evolutionary psychologists and game-theory economists to work together. I spent months talking to researchers individually, in groups, and at conferences. It was the most frustrating experience of my professional life, for we psychologists just did not understand the economists, and they did not understand us. We were interested in real people; they were interested in idealized markets. We liked experiments; they liked proving mathematical theorems. We published ideas about human nature; they published results about Pareto-dominant equilibrium selection in mixed-motive games (don't ask).

My crisis point came at a 1999 conference that I organized in London on the origins of people's economic preferences. We psychologists thought that economists would enjoy hearing about our preference experiments, so that they could develop more accurate and sophisticated models of human economic behavior. How wrong we were. It became clear that economists still followed a "revealed preferences" doctrine, which holds that consumer preferences are psychological abstractions—hidden, hypothetical states that cannot be measured or explained apart from the purchases that they cause. If preferences are revealed only through purchases, and not through questionnaires, interviews, or focus groups, then it is redundant to study preferences apart from actual consumer spending patterns, to speculate about the origins of preferences, or to conduct market research on preferences for hypothetical products. In short, the revealed preferences doctrine suggests that psychology is irrelevant to economics. (This was before the psychologist Daniel Kahneman got the 2002 Nobel Prize in Economics for his work on decisions and preferences.) So, the economists gradually drifted away from the conference, leaving the psychologists to nurse our bruised egos, in the company of some strange-looking folks we hadn't seen before.

These folks weren't like the academics at the conference. They were forty-five but looked twenty-five; they had funky clothes and hair; they spoke with torrential enthusiasm; they gave out business cards with baffling job titles (Cool-Hunter, Frenzy-Mistress, Vice-President of Buzz, Meme-Seeder). They were the marketers, and they were hot for psychology. They actually cared about people's preferences—where they come from, how they worked, and how to profit from them. I talked for hours with them, and a new world opened up.

Over the next several years, I read everything I could about marketing, advertising, public relations, market research, product design, retailing, branding, positioning, and consumer behavior. It felt as if my latent interest-in-business genes had finally turned on. (My maternal grandfather, Henry G. Baker, had been a professor of management and marketing at the University of Cincinnati, and most of his five sons now run private equity funds.) I taught courses on the evolutionary psychology of consumer behavior, first to undergraduates at UCLA in 2000 as a visiting professor, then at UNM to graduate students. I became fascinated by incisive portrayals of the consumerist lifestyle in movies such as *The Matrix, Existenz, American Beauty,* and *Idiocracy,* and in novels by Chuck Palahniuk, Douglas Coupland, Nicholson

Baker, and J. G. Ballard. I talked about marketing with everybody I knew who was involved in it—old high school friends, relatives, neighbors, local business school faculty. I subscribed at various times over the past seven years to any periodical that seemed likely to reveal something new about consumerism: *Architectural Digest, AutoWeek,* the *Baffler,* the *Chronicle of Higher Education, Consumer Reports,* the *Economist, Gourmet, Harper's, Maxim, Men's Fitness, Money, PC Gamer, Premiere, Rolling Stone, Stuff, Wired, Worth,* the *Utne Reader,* and *Vanity Fair.* I also pulled articles and ads from the occasional issue of *Action Pursuit Games, Adult Video News, All About Beer, Atomic Ranch, Christian Music Planet, Cosmetic Surgery Times, Frozen Food Age, Guns & Ammo, Hooked on Crochet! Hot Boat, Log Home Living, Luxury SpaFinder, Meat Processing, Modern Bride, Modern Dog, Monster Muscle, New Age Retailer, Packaging Digest, Pet Product News, Sport Compact Car,* and *Tropical Fish Hobbyist.* This was not always as fun as it sounds. I also read a few hundred books on consumerism and business, in search of good ideas.

I started to see that marketing underlies everything in modern human culture in the same way that evolution underlies everything in human nature. Writers have agents; movies have publicists; politicians have press secretaries. Magazines are published not to inform readers, but to sell the market segment—the readers' attention—to advertisers. Almost nothing in popular culture gets there by chance or gossip, by the unsupervised spread of memes from one mind to another. Everything has been put on the public's radar screen deliberately by marketing professionals of one sort or another.

I realized, in short, that if you weren't tuned in to marketing, you were missing the elephant in culture's living room.

# 3

# Why Marketing Is Central to Culture

MARKETING IS NOT just one of the most important ideas in business. It has become the most dominant force in human culture. If this sounds like an outrageously strong claim that no rational person could believe, consider that much of this disbelief stems from misunderstanding "marketing" as a pretentious term for advertising. But marketing is far more than that. It is, ideally, a systematic attempt to fulfill human desires by producing goods and services that people will buy. It is where the wild frontiers of human nature meet the wild powers of technology. Like chivalrous lovers, the best marketing-oriented companies help us discover desires we never knew we had, and ways of fulfilling them we never imagined.

Almost everything we can buy has been shaped by some marketing people in some company thinking hard about how to sell us things that we think will make us happier. Adam Smith's "invisible hand" has spawned the invisible eye. Production is no longer guided by the clumsy feedback provided by last quarter's profit figures, but by empirical research into human preferences and personalities: focus groups, questionnaires, beta testing, social surveys, and demographics. Psychology has given way to market research as the most important investigator of human nature. For example, as of 2004, about 212,000 Americans worked as market and survey researchers, whereas only about 37,000 worked as psychology professors.

Markets themselves are ancient, but the concept of marketing in its modern form arose only in the twentieth century. In agricultural and mercantile societies there were producers, guilds, traders, bankers, and retailers, but economic consciousness was focused on making money, not on researching and fulfilling consumer desires in any systematic way. Only through trial and error did Albrecht Dürer learn what kinds of prints would sell, or Thomas Chippendale learn which of his chairs would prove fashionable. With the Industrial Revolution, mass production led to a greater emphasis on the cost efficiency of production rather than the satisfaction of the customer. As markets matured in the early twentieth century, businesses had to compete harder for market share, but they did so through advertising and sales promotions aimed at unloading their goods on resistant customers.

Only gradually did corporations understand the relevance of psychology to sales. A key figure was Edward Bernays (1891–1995), a founding theorist of propaganda, public relations, and advertising. Bernays was Sigmund Freud's nephew, and used psychoanalytic insights to address what he called the problem of "engineering consent" in a democratic society. He consulted on ad campaigns for Dodge, Procter and Gamble, General Electric, and Cartier, and helped the United Fruit Company (now known as Chiquita) overthrow the Guatemalan government in 1954. In his 1928 book *Propaganda*, Bernays argued

> The conscious and intelligent manipulation of the organized habits and opinions of the masses is an important element in democratic society. Those who manipulate this unseen mechanism of society constitute an invisible government which is the true ruling power of our country.

Yet, even Bernays realized that effective manipulation of public opinion required listening to the beliefs and desires of consumers and citizens. Governments and corporations need to listen in confessionals, not just to shout from the pulpit. Good public relations require good public opinion polls, not just good propaganda.

By the time Willie Loman was lamenting the fall of traditional hucksterism in *Death of a Salesman* in 1949, several consumer-goods companies had developed a more respectful, inquisitive attitude toward the consumer. The marketing revolution they launched came with the same sense of wondrous inevitability that accompanies all scientific revolutions. That a company should produce what people desire, instead of trying to convince people to buy what the company happens to make, was a radical idea that seems obvious only in retrospect. These corporations established marketing departments dedicated to finding out what people want from their detergents, soaps, and lightbulbs. Their success spawned imitators, and almost all large corporations now include marketing arms that are supposed to coordinate all aspects of product research and development, advertising, promotion, and distribution.

By the 1960s, as more and more marketing executives were promoted to CEO positions, some firms adopted the modern "marketing orientation," in which everything the firm does is aimed at making profits by satisfying consumers. This constituted an invisible revolution in the 1960s, and though it did not get the same press as the sexual revolution, the hippies, or the New Left, unlike these counterculture trends, the marketing revolution radically changed the way business works. (Indeed, the marketing revolution was largely responsible for the popular spread of countercultural values through cool new products like the Volkswagen T2a Bus, the Enovid contraceptive pill, and Jimi Hendrix records, all of which were often combined enjoyably.)

While the marketing orientation has become commonplace in companies that produce things for individual customers, such as clothes, cars, televisions, and movies, it remains rare in the heavy industries (steel, coal, oil, and paper), where the immediate consumers are other businesses, and where conspicuous consumption and luxury branding are less important. The marketing orientation is also still poorly developed in most service industries such as banking, law, government, the police, the military, medicine, charity, and science. Indeed, most leaders in these sectors do not think of themselves as working in service industries. But until they do, they will not bother using market research to shape their services to their customers' desires, and their institutions will lose market share to those that do.

The transition from the production orientation to the marketing orientation is still under way, and remains one of the most important but least understood revolutions in human history, marking a decisive power shift from institutions to individuals. Production made workers into technology's servants; marketing, ideally, makes consumers into technology's masters. Marketing zealots might even take the view that the marketing revolution renders most of Marx irrelevant: What meaning could "alienation" and "exploitation" have when businesses work so hard to fulfill our desires as consumers?

Generally speaking, intellectuals still don't understand marketing. It is largely invisible to right-wing economists, who think prices carry all the information about supply and demand that markets need in order to produce the goods and services that people want. There was no role for market research in the worldviews of the economists Adam Smith, Friedrich Hayek, Milton Friedman, or Gary Becker. To left-wing social scientists, journalists, and Hollywood scriptwriters, in contrast, marketing means nothing more than manipulative advertising by greedy corporations. Since they rarely deign to talk to businesspeople, they believe that modern business works like the evil Omni Consumer Products corporation from *Robocop*. The rare professors who acquire some modest net worth tend to learn much more about investments than about marketing, because while investment advice is everywhere (CNBC, Fox Business Network, personal finance magazines), marketing knowledge lurks as a sort of arcane magic behind this financial-product hucksterism.

One problem is that marketers, like all professionals and academics, are prone to showing off their expertise by using distinctive terms and concepts that are baffling to eavesdroppers. When disempowered subcultures use private jargon, they sound cute. But when marketers with heavy economic firepower do so, they can sound both hilarious and necromantic, like Pentagon acronyms. Consider these lines overhead at the 2006 Intelligent Printing and Packaging Conference, posted on one of Bruce Sterling's blogs:

- "It's our metallo-organic approach versus the incumbent technologies"

- "Thermochromic ink is the Pet Rock ink of the New Millennium"
- "We need a taxonomy for printing-that-is-no-longer-printing"
- "Electronic cardboard blurs the line between printed objects and the virtual world"
- "It's bubble, bubble, toil, and trouble in conductive polymers"

Such canny observations no doubt mean something, but it's not clear what.

Even within business, although most younger managers understand marketing at a practical level, they do not know how to talk about it as a cultural, economic, social, and psychological revolution, as it is not presented that way to them in business school. Business journalists, likewise, have not brought the marketing revolution into public discourse the way they have brought the "New Economy" of the Internet to the public's attention. Pundits still talk as if we are moving from an industrial era based on mass production to an information era based on mass entertainment.

Like fish unaware of water, we do not realize that we live in the Age of Marketing. It does not much matter whether products are material or cultural, sold in stores or online. What matters is that products are systematically conceived, designed, tested, produced, and distributed based on the preferences of consumers rather than on the convenience of producers. The New Economy, "Web 2.0," and "social network marketing" are just the most recent stages in the marketing revolution.

How can we understand this revolution? There are two analogies from history that can help us think about it. Democracy can be seen as the marketing concept applied to government. The American and French revolutions brought the marketing concept to politics long before it gained a toehold in business. The production-oriented state asked what taxpayers could do for it; the marketing-oriented state asks what it can do for voters. Citizens demanded the vote so they could tell government what state services they wanted long before consumer focus groups were telling manufacturers what goods they wanted. "No taxation without representation" came long before "No profits without market research."

Even before these political revolutions, the Protestant Reformation applied the marketing insight to religion. Martin Luther and John Calvin organized churches to fill the emotional needs of worshipers, not the fiscal interests of priests. They were dissatisfied with a production-oriented papacy that churned out costly rituals in a dead language within opulent cathedrals. They crafted a new form of Christianity based on local languages, simple churches, and glorious music. The thirty thousand current denominations of the Christian faith are just what we would expect from efficient market segmentation given diverse consumers of religious services. Similar shifts occurred from production-oriented Hinayana Buddhism to market-oriented Mahayana Buddhism, and from Orthodox to Reform Judaism. The common denominator in business marketing, political democracy, and religious reform is the transfer of power from service providers to service consumers.

Is the marketing revolution a good thing? On the upside, it promises a golden age in which social institutions and markets are systematically organized on the basis of strong empirical research to maximize human happiness. What science did for perception, marketing promises to do for production: it tests intuition and insight against empirical fact. Market research uses mostly the same empirical tools as experimental psychology, but with larger research budgets, better-defined questions, more representative samples of people, and more social impact. Ideally, marketing's empiricism works like Rogerian psychotherapy, in which the therapist restates and reflects the patient's concerns. Marketing holds up a mirror to our selves, reflecting our beliefs and desires so we can recognize, remember, evaluate, and transform them. The invention of real mirrors empowered people to accept or reject potential modifications to their appearance with greater accuracy and objectivity, allowing them to try different makeup, hairstyles, and fashions, to judge what looks good. The marketing revolution empowers us in a similar way, on a longer timescale. It allows us to accept or reject potential ways of displaying our traits through our product choices. We can try on different lifestyles, experience the results, and perhaps even change our consumer preferences if we're dissatisfied.

On the downside, marketing is the Buddha's worst nightmare. It is the grand illusion, the Veil of Maya, turned pseudoscientific and backed by billion-dollar advertising campaigns. It perpetuates the delusion that desire leads to fulfillment. It is the enemy of mindful human consciousness, because consciousness is content with its own company, and needs little from the world.

The trouble is not that marketing promotes materialism. Quite the opposite. It promotes a narcissistic pseudospiritualism based on subjective pleasure, social status, romance, and lifestyle, as a product's mental associations become more important than its actual physical qualities. This is the whole point of advertising and branding—to create associations between a product and the aspirations of the consumer, so the product seems to be worth more to the consumer than its mere physical form could possibly warrant. Marketing actually avoids materialism at all costs, for if consumers comparison shopped solely on the basis of objective material features and costs, the products themselves would be reduced to commodities—and commodities cannot be sold for serious profits in a competitive market.

For example, tap water (about $0.006 per gallon in Albuquerque) is a low-profit commodity, whereas Glacéau SmartWater ($1.39 per thirty-four-ounce bottle, or $5.20 per gallon) is a high-profit branded product. SmartWater sounds like a magical intelligence-boosting elixir from the French Alps, so it can be sold for 870 times the price per volume of commodity water, even though it is really just distilled water with some added electrolytes (a bit of calcium chloride from limestone and magnesium chloride from seawater). However, right after Coca-Cola acquired Glacéau in 2007 for $4.1 billion, SmartWater started to be advertised with the image of a nearly nude Jennifer Aniston. So, commodity water plus limestone and seawater and a nice bottle, plus Aniston's beauty and fame, yields a profitable brand.

Thus, a world run by marketing to profit from consumer desires will never allow itself to be "commodified" into a "materialistic" world. Rather, it could easily transmute into a virtual reality where neither products nor consumers require any physical qualities at all. Marketing's logical culmination would not be crass materialism, but the seductive immateralism of *The Matrix* or *Second Life*.

Marketing also creates some more immediate problems. Like democracy, it forces intellectual and cultural elites to confront their patronizing attitudes toward the masses. Elites do not always like companies and states that provide what the people desire. Consumers may want sweets, fats, and sugars; cigarettes, beer, and marijuana; motorcycles and handguns; porn videos and prostitutes; breast implants and Viagra; reality TV and formulaic anime. Likewise, if everybody voted, they might want the death penalty, prayer in schools, book burning, ethnic cleansing, fascism, and *American Idol*. Plato clearly saw the difference between a mass democracy based on universal suffrage, and a republic based on the utopian visions of elites. For the elite, marketing's populism can be an alarming prospect. Plato thus rejected the marketing orientation—including democracy, which is marketing applied to politics—as a basis for social organization. His ideal benevolent dictator, the philosopher-king, does not organize focus groups, conduct market surveys, or hold elections to decide his policies. The common folk cannot be trusted to understand their true long-term interests, for the mismatch between their primitive instincts and the behavioral demands of civilized life are so severe that the enlightened minority must control the ignorant majority, for the greater good. Confucius had similar views: the patriarch must rule the family, just as the emperor must rule the nation, to impose civilized order on natural anarchy.

This Platonic-Confucian tradition dominated European and Asian political theory for millennia. It can still be seen today, whenever elites argue that the state should collect taxes and provide certain services that people can't or won't buy as individuals. Sometimes these state-organized services seem sensible (roads, fire departments, health care, the BBC), sometimes not (farming subsidies, fraudulent wars, bridges to nowhere). The Platonic-Confucian ideal also comes into play whenever elites argue that some product or behavior must be banned. (Sometimes the elites have a point: even among Second Amendment extremists, few would advocate that your local Target store should be allowed to sell FIM-92 Stinger surface-to-air missiles.)

Marketing, like democracy, has the (often untapped) potential to be anti-arrogance, anti-power, and anti-idealism. It can, in principle, replace elitist progressive visions based on the illusion of popular consent with the reality of a world shaped to fulfill ordinary human desires. It is tempting to minimize the marketing revolution, to naïvely propose that the most significant revolutions of the past millennia have been technological inventions that expand production capabilities, or scientific ideas that inform elite ideals. If we do choose to ignore the marketing revolution, we do so because we are terrified of a world in which our elite ideals lose their power to control the fruits of technology. (If you have the leisure time, education, and inclination to read this book, you are obviously a member of the elite.) Marketing threatens to put infinite production ability in the service of infinite human lust, gluttony, sloth, wrath, greed, envy, and pride. It portends a world of *Idiocracy*, Cinnabons, and Super Bowls. It threatens to atomize human society into 6 billion navel-gazing blog writers.

Or is the elite's fear of that prospect just another self-deceptive rationale for keeping a stranglehold on power? Fear of an economy based on market research, like Plato's fear of democracy based on universal suffrage, is based on contempt for fellow members of our species. Elites hate to recognize the marketing revolution because they hate to acknowledge that contempt. Marketing is the most important invention of the past two millennia because it is the only revolution that has ever succeeded in bringing real economic power to the people. It is not just the power to redistribute wealth, to split the social cake into different pieces. Rather, it is the power to make our means of production transform the natural world into a playground for human passions.

Ecologists estimate that humans now consume more than half

our planet's "net primary productivity"—more than half the biomass grown each year on earth. One lucky species, out of 20 million, sucks up half of the biosphere's annual output, and transforms it into work roles and leisure activities that are structured mainly by marketing. Marketing does not just dominate human culture; since human culture dominates the matter and energy flows that constitute terrestrial life, it also, at this historical moment, dominates life on earth.

## Marketing Versus Memes

This blindness to marketing's cultural role became especially clear when I joined a debate about memes, moderated by Richard Dawkins, at Oxford University in May 1999. My debating partner, the British psychologist Susan Blackmore, had just published her book *The Meme Machine*. She argued, following Dawkins, that much of human culture reflects an evolutionary competition between memes: information units such as stories, anecdotes, ideas, catchphrases, or jingles that can be remembered and repeated to others. Memes that are salient, memorable, and communicable (like celebrity gossip and human interest stories) are expected to proliferate and spread. Memes that are irrelevant and forgettable (like the fact that a proton has about 1,836 times the mass of an electron) should fade quickly from popular consciousness (despite the best efforts of high school physics teachers). According to Blackmore, human popular culture consists of successful memes that reflect the interests and preferences of individual humans.

The meme idea has always seemed fascinating and provocative, especially in Blackmore's book. However, I argued a somewhat different line: most successful memes are imposed top down by marketing, in the interests of certain powerful individuals, groups, and institutions. It seemed clear that the most successful memes—religions, political ideologies, languages, cultural norms, technologies–have been disseminated by churches, states, school systems, and corporations with immense wealth and power. In principle, marketing responds to pre-existing consumer preferences. In fact, marketers sometimes refer to their work as "cultural engineering"—the intentional creation and dissemination of new culture units (memes) through advertising, branding, and public relations.

Even the proliferation of ordinary memes (such as buzz about films, new social and political issues, countries to fear this year) is dominated by the six global media conglomerates:

- TimeWarner ($45 billion in revenue, 87,000 employees as of 2006), including Warner Bros., New Line Cinema, AOL, CompuServe, Atlantic Records, HBO, CNN, Time Warner Cable, Turner Broadcasting, Time-Life Books, and the magazines *Time, Life, Money,* and *People*
- Disney ($34 billion in revenue, 133,000 employees), including Touchstone, Miramax, Buena Vista, ABC TV, ESPN TV, Hyperion Books, *Discover* magazine, and ABC Radio Networks
- NewsCorp ($25 billion in revenue, 47,000 employees), including Twentieth Century Fox, Fox TV, Sky satellite TV, Sky Radio, HarperCollins Books, *TV Guide,* and 175 newspapers
- Vivendi Universal ($20 billion in revenue, 34,000 employees), including Universal Studios, Geffen Records, Polygram, Universal Music Group, Canal+ TV, and Universal Television Group
- Bertelsmann ($20 billion in revenue, 97,000 employees), including UFA Film and TV, Barnes and Noble, BMG Music Publishing, RCA Records, AOL Europe, and the publishers Ballantine, Bantam, Crown, Doubleday, Dell, Fodors, Knopf, and Random House
- Viacom ($10 billion in revenue, 9,500 employees), including Paramount, United Cinemas, CBS TV, MTV, Showtime TV, Simon & Schuster, Infinity Radio, and Viacom Outdoor advertising

These conglomerates relentlessly cross-promote their TV channels, films, magazines, and books through all available media. For example, if Warner Bros. releases a big-budget film such as *The Dark Knight*, it will typically be featured on the covers of *Time* and *People* magazines, reviewed favorably on CNN, and well advertised on AOL. This is not conspiracy theory; it's just good business sense and standard operating procedure for media conglomerates.

Apart from these big six media conglomerates, there are the four big advertising holding companies:

- Omnicon ($13 billion, 61,000 employees)
- WPP ($12 billion in revenue, 100,000 employees in 106 countries as of 2007)
- Interpublic ($7 billion, 43,000 employees)
- Publicis ($6 billion, 44,000 employees)

Most academics have never heard of these companies, but they are at the heart of cultural engineering, as they are involved not only in advertising, but also design, marketing, media buying, public relations, and lobbying. They design the memes, buy the airtime and column inches to distribute them, and measure how well the memes are achieving their purposes in promoting consumer, investor, and political recognition for their clients. Altogether, about $400 billion per year is spent in the global ad market—money spent specifically to promote some memes, brands, products, and people at the expense of others.

Consider another example of cultural engineering: food preferences. Every evolutionary psychology textbook suggests that our fast-food cravings for fat, salt, and sugar are innate, evolved preferences. The theory is that because these nutrients were so rare and valuable in prehistory, we inherited an insatiable desire for them that is now counterproductive, making us fat and sick. Honey was hard to get in the Pleistocene, so now we can't help but eat two-hundred-calorie Krispy Kreme doughnuts. This evolutionary view nicely accounts for some cross-cultural universals in food preferences.

Meme theory offers a different view: maybe we consume steaks, doughnuts, and sodas because we've seen others do so, and have imitated their eating habits. We could just as well have ended up favoring pickled tofu and Siberian kale, but the random dynamics of meme evolution took us in a different direction. This meme view might likewise explain some cross-cultural differences in food preferences, such as why Americans tend to turn all naturally savory foods into sweet desserts, by such measures as adding barbecue sauce to meat, ketchup to french fries, honey-mustard dressing to salad, sugar to bread, and corn syrup to water ("soda").

While their insights are valuable, it would be helpful for both evolutionary psychologists and meme theorists also to recognize the economic, political, and marketing power of the global food industry. In the United States, we have a lot of fat, salt, and sugar in processed foods partly because there are rich, powerful trade organizations that lobby politicians very effectively for government subsidies and contracts, weaker regulations, and tort reform to minimize liability, including the National Council of Chain Restaurants, National Grocers Association, Food Products Association, Food Marketing Institute, and Grocery Manufacturers of America. The National Restaurant Association represents the nine hundred thousand U.S. restaurants, which employ 12.2 million workers and earn $476 billion in revenue per year. The National Cattlemen's Beef Association represents eight hundred thousand ranchers who "harvest" about 26 billion pounds of beef per year from 35 million cattle. The National Chicken Council represents massive companies such as Tyson, Gold Kist, Pilgrim's Pride, and ConAgra, which sell about 600 million pounds of chicken flesh per week in the United States, by killing about 8 billion chickens per year. Further promoting fat and protein consumption are the American Meat Institute, National Pork Board, National Turkey Federation, International Dairy Foods Association, and National Milk Producers Federation. To promote salt consumption, we have the Snack Food Association and National Association of Convenience Stores. To promote sugar consumption, we have the Sugar Association, Association for Dressings and Sauces, and International Jelly and Preserves Association. The Corn Refiners Association is especially important, since it represents the U.S. "corn wet-milling industry," which manufactures about 25 billion pounds of high-fructose corn syrup per year. Corn syrup is the main ingredient in sodas (apart from water), and each American consumes on average about forty-five grams of it per day.

So, we doubtlessly have some innate liking for fat, salt, and sugar. The lobbyists and trade organizations do not create the demand for these tastes out of thin air—otherwise the Pickled Tofu Marketing Institute and U.S. Kale Association would have greater funding, influence, and success. Nonetheless, these more-powerful industry groups hugely amplify our evolved food preferences through massive political clout and marketing budgets for their food groups.

The shaping of such ideas, tastes, norms, habits, and memes by social power systems is exactly what the social sciences study. It is the very lifeblood of political science, sociology, and media studies. These sciences have realized through decades of research that they can't jump straight from individual psychology to mass culture through a simple model of meme evolution; that would be as naïve as market fanatics thinking that political anarchy plus the economics of supply and demand would yield utopia. We also have to consider social institutions and interests. Meme perpetration—conscious, deliberate, institutionalized strategies for shaping popular views and preferences—is what millions of people are paid to do every day when they work as marketers, advertisers, retailers, or public relations experts.

The market fanatics are right in one respect: marketing's power is quite decentralized. There is no unified conspiracy, no secret Masonic Temple, to perpetuate capitalism, consumerism, patriarchy, heterosexism, racism, or general mass stupidity and apathy. The World Trade Organization is just 630 folks working in a five-story office building at 154 Rue de Lausanne in Geneva. Mostly, marketers are not trying to perpetuate the power systems that social scientists analyze; they're just trying to increase market share for their companies. Marketers are often portrayed as evil geniuses, but in reality, they're typically floundering around like everybody else. They try to keep up with the latest consumer psychology fads by reading the shortest available pop-business books by eccentric writers with extremely large or small quantities of hair.

So, none of the extreme views offered by modern science works very well to understand marketing. The innate-preferences theory and meme theory neglect marketing power entirely; the social science conspiracy theories neglect the decentralized, dog-eat-dog, ill-educated competition among marketers. As a result, most behavioral sciences—psychology, anthropology, sociology, economics, political science—have rarely taken marketing seriously, with the result that they have largely ignored the mainspring of modern culture, the central force that amplifies, dampens, distorts, frustrates, or fulfills human nature.

# 4

# This Is Your Brain on Money

WHILE MARKETING is central to modern culture, the consumerist mind-set is central to marketing. To understand this mind-set objectively, we have to reach escape velocity from its gravity well, so that we can turn around and examine it from a distance. This is hard to do when one's ego and self-esteem are wrapped up in one's identity as a consumer. To detach from consumerism, it may help to feel embarrassed, alienated, and betrayed by it. Like a cirrhotic drunk at his first Alcoholics Anonymous meeting, we may find that it helps to admit that our life choices have had a whiff of insanity about them.

Given that perspective, what sort of mental illness is most analogous to consumerism? Is it, for example, most like depression, or schizophrenia, or post-traumatic stress disorder? I think the relevant comparison here is to narcissism, technically called narcissistic personality disorder. Personality disorders in general are deeply ingrained, lifelong, pervasive, maladaptive problems in living and relating to others. Narcissism in particular is a pervasive pattern of self-centered, egotistical behavior that usually begins by early adulthood, and that combines an intense need for admiration by others with a lack of empathy for others.

## Narcissism and Consumerism

Freud introduced the concept of narcissism in 1914, drawing on the ancient Greek myth of Narcissus—a handsome, seductive young man who rejected the love of the wood nymph Echo, and fell in love instead with his own reflection in a pool of water, eventually withering away and turning into the flower that bears his name. Narcissism is thus love of one's external image as it would be loved by another—combined with contempt for others who actually feel love for one's inner qualities. The best archetype for consumer narcissism is probably Paris Hilton's self-branded fragrance, Just Me (that is, Not You and Not We). It's the scent of solipsism.

The key diagnostic features of narcissism (according to psychiatry's bible, the cumbersomely named *Diagnostic and Statistical Manual of Mental Disorders,* fourth edition, text revision) are:

- selfishness (taking advantage of others, lacking empathy)
- arrogance (haughty, contemptuous attitudes, plus rage when frustrated or contradicted)
- exceptionalism (belief that one is special and can only be appreciated by other high-status people)
- sense of entitlement (expecting special treatment and automatic compliance with one's wishes)
- admiration seeking (needing excessive attention, affirmation, praise, and deference)
- success fantasizing (obsessive ambitions about unlimited success, power, brilliance, beauty, sexual power, or ideal love)
- grandiosity (exaggerating one's talents, achievements, and status)
- victim mentality (blaming the outside world for one's failures and disappointments)
- anhedonia (inability to enjoy simple pleasures)
- emotional instability (when cut off from the "narcissistic supply" of adulation from others, narcissists feel sad, hopeless, and even suicidal)

These core symptoms lead narcissists to view themselves as stars in their own life stories, protagonists in their own epics, with everyone else a minor character. (They're like bloggers that way.) They talk about their lives, careers, and families as if nobody else were in the picture. Their senses are numb, so they seek ever more intense stimulation.

They feel irritable and show a low frustration tolerance. They some-times reward themselves with impulsive, hedonistic extremes: alcohol, drugs, gambling, binge eating, shopping on credit, promiscuous sex. They sometimes perceive a "grandiosity gap" between their inflated self-esteem and their actual accomplishments, leading to an unstable sense of self-worth and periodic self-doubts and depression. Narcis-sists tend to seek pleasure not through informal social interaction with equal-status others, but through self-stimulation (fiction reading, TV watching, drug taking, masturbating) and showing off (ritualized displays to lower-status admirers, as in wearing absurdly impractical fashions, or throwing extravagantly destructive parties). Tech-savvy narcissists are also likely to do a lot of "ego surfing" (googling their own names to see what comes up), and "blog streaking" (revealing overly personal details in their blogs).

Does that description call to mind anyone you know? If you're a mature adult, it may sound like all young adults. If you're from a poor country, it may sound like all Americans. If you're a woman, it may sound like all men. Most of us can seem narcissistic to some of the people some of the time. However, true narcissistic personality disorder—extreme, relentless, hard-core narcissism—is estimated to affect only about 1 percent of the population. Yet the capacity for narcissism, under certain conditions, seems present in most ordinary humans. Runaway consumerism works largely through creating these conditions and tapping this capacity.

Narcissists rarely think anything is wrong with them, and there-fore don't seek treatment. If they do show up at a therapist's, it's usu-ally because a long-suffering spouse has issued an ultimatum. Even in treatment, they rarely get better. Giving them Prozac, self-help books, or self-esteem-boosting exercises just tends to increase their grandi-osity, sense of entitlement, and victim mentality. Therapy feeds their egos, often becoming just another source of "narcissistic supply" (self-focused attention and praise). Similarly, unself-conscious consumers rarely think anything is wrong with them. They're probably married to another unself-conscious consumer, so they don't even have the spousal pressure to seek treatment, insight, or change.

We don't yet know what genes, environmental triggers, and random events during brain growth lead some people to develop true narcissism. But once we do, it seems likely that those factors will overlap with the factors that promote runaway consumerism. The environmental triggers for both seem likely to include the relentless self-esteem-boosting messages sent to children of certain countries by their parents and teachers. Kids told fifty times a day that they have "done awesome," regardless of their talents and virtues, seem likely to acquire a grandiose sense of entitlement and a penchant for egotistical self-indulgence—not to mention an inability to use adverbs properly.

## The Two Faces of Consumerist Narcissism

Narcissists tend to alternate between public status seeking and private pleasure seeking. I believe that these two faces of narcissism are also the two key components of the consumerist mind-set. We buy things for status or for hedonism, to show off to others or to please ourselves, to send fake fitness indicators to others or fake fitness cues to ourselves. Remember that fitness indicators are signals of one individual's traits and qualities (good genes, good health, good social intelligence, and so on) that are perceivable by others—signals like the peacock's tail, the bowerbird's bower, or the consumer's iPod. Generally, animals have no conscious awareness that, by displaying fitness indicators, they will attract mates, friends, and help from kin; they simply feel an urge to do the displays under certain conditions, and they reap the evolutionary benefits.

Fitness cues, in contrast, are features of an individual's environment that convey useful information about local fitness opportunities—ways to increase one's survival chances or reproductive success. Darkness is a cue for danger (reduced survival chances), so it induces fear and shelter seeking. For predators, the scent of prey is a cue for food (increased survival chances), so it motivates pursuit, attack, and ingestion. For males, the cues that identify fertile females of their own species carry information about mating opportunities (increased reproductive success), so they motivate pursuit, courtship, and copulation. Our perceptual systems have evolved to pay the most attention to these sorts of fitness cues, because, in evolutionary terms, they are the only things worth noticing about one's world. (Natural selection cannot favor animals' responding to any cues that do not identify an opportunity to promote their survival or reproduction.) Further, animals evolve motivation systems to surround themselves with positive, fitness-promoting cues (which evolve to "feel good"), and to avoid negative, fitness-threatening cues (which evolve to "feel bad"). At the evolutionary level, animals are always under selection to survive and reproduce. But at the subjective level, they are always motivated to chase the fitness cues that feel good—not because they consciously understand that natural pleasures are associated with evolutionary success, but because they have been shaped to act as if they understood that association unconsciously.

So, there is a parallel between nonhuman animals displaying fitness indicators and chasing fitness cues, and humans seeking status and pleasure, whether as narcissists or consumers. This does not mean that all consumers should be clinically diagnosed with narcissism, but rather, that all human brains have a deep and abiding interest in two big sets of evolutionary goals: displaying fitness indicators that were associated with higher social and sexual status in prehistory, and chasing fitness cues that were associated with better survival, social, sexual, and parental prospects in prehistory. These two branches of universal human nature spontaneously grow large and malignant in the 1 percent of us who are true narcissists. These two branches are also watered, fertilized, and hot-housed by consumerist capitalism. So, narcissism and consumerism are two related ways in which the drives for displaying fitness indicators and chasing fitness cues can take over our lives—often to the exclusion of empathy, intimacy, friendship, kinship, parental responsibility, and community spirit.

## The Two Faces of the iPod

Almost all advertisements appeal to status seeking, or pleasure seeking, or both. An excellent example is the sixth-generation iPod Classic, released in 2007. It is a "portable media center" that fits in one hand; it is 4 by 2.4 inches in size, about half an inch thick, and weighs about six ounces. Yet its aluminum case contains 160 GB of storage, which can hold forty thousand songs, twenty-five thousand photos, or two hundred hours of video. Its 2.5-inch, 640-by-480-pixel color screen can display MPEG-4 movies, TV shows, and games. It can play music, podcasts, or audiobooks for about forty hours on one charge of its lithium ion battery. It retails for about $350. However, if you actually filled its hard drive completely with $0.99 songs downloaded from iTunes, the iPod's total cost would be $40,000.

The iPod is not a typical example of "conspicuous consumption" in Thorstein Veblen's sense, as it is an affordable little device that sells by the millions. However, it does demonstrate that the two aspects of consumer narcissism are often at work even in products that are not ostentatious or exclusive. First, iPods display coolness, status, and wealth through their sleek design, brand recognition, and moderately high cost (relative to a typical teenager's allowance). You can customize their appearance by uploading new screen images, covering the click-wheel interface with "wheel art," and covering the whole case in "wraps," "bands," or "skins" in any color or texture of silicone, or in leather cases. The distinctive iPod earbuds can be swapped out for more exotic-looking Earpollution D33 Earbuds (showing a radiation hazard symbol, $13), the SkullCandy Full Metal Jackets (showing a skull, $50), or the Heyerdahl iDiamond ears (each earbud covered in 204 diamonds; $6,400).

At the same time, iPods embody the self-stimulation aspect of narcissism. They play music that no one else can hear, or videos that no one else can see. They are pleasure-delivery systems, private mediascapes. They encourage a narcissistic worldview in which iPod users are the stars of their own action-romance epic, with their own

subjective soundtrack to drown out the irritating voices of the minor
characters around them. (The fact that those minor characters must
endure the thumpy-thumpy bass leaking from the Earpollution Earbuds is of no concern.)

The following table explores these two main aspects of consumer
narcissism in more detail.

### THE TWO FACES OF CONSUMERIST NARCISSISM

|  | showing off | self-stimulating |
| --- | --- | --- |
| *Basic functions* | trait display | pleasure delivery |
| *Intended audience* | others | self |
| *Life goals* | success, fame, fortune | happiness, fun, fulfillment |
| *Narcissism symptoms* | grandiosity | solipsism |
|  | admiration seeking | pleasure seeking |
|  | obsessive status fantasies | obsessive self-stimulation |
|  | arrogance, ambition | perfectionism, irritability |
|  | lack of humility | lack of empathy |
| *Associated deadly sins* | pride, avarice, envy | lust, gluttony, sloth, wrath |
| *Typical activity* | work, socializing | leisure, dreaming |
| *Typical food* | Kobe beef, foie gras | lamb vindaloo, mango kulfi |
| *Typical drink* | rare burgundy, Red Bull | hot chocolate, margarita |
| *Typical clothing* | business suit | lingerie |
| *Typical house feature* | entry hall, dining room | media room, master bath |
| *Typical software product* | personal Web page | computer game |

|  | showing off | self-stimulating |
|---|---|---|
| *Typical college major* | finance, premed biology | literature, psychology |
| *Typical reading material* | quotable nonfiction | escapist fiction |
| *Typical film genre* | foreign, classic | action, porn |
| *iPod features* | sleek design | hard drive size |
|  |  | sound & screen quality |
|  | Apple | battery life, |
|  | branding | light weight, |
|  |  | custom covers |

## Showing Off

*Spent* focuses on the showing-off forms of consumer narcissism represented in the left column. (Gad Saad's 2007 book, *The Evolutionary Bases of Consumption,* dealt a bit more with the right column.) As we will see, a surprisingly high proportion of products are designed and marketed for showing off—as narcissism projectors, trait amplifiers, fitness indicators, signals of health, wealth, or virtue. This has been well understood by every intelligent observer of capitalism since Adam Smith, including Thorstein Veblen, Vance Packard, and Robert H. Frank.

Yet we routinely choose to view consumer narcissism as something that other people did in the historical past, or that they do in other cultures and subcultures. We rarely have clear insight into our own forms of consumer narcissism, a fact that is especially true for educated citizens in developed economies—such as most readers of books like *Spent.* We may feel contempt for the forms of conspicuous consumption that we consider crass, outré, and infra dig: the Botox, Hummers, and McMansions of the aesthetically misguided nouveaux riches. At the same time, we frame our own less-conspicuous varieties of consumer narcissism as natural, reputable, enlightened forms of authentic self-expression, merited achievement, and civic virtue. One can hang a Hampshire College degree (a signal of liberal countercultural openness that cost $171,540 for four years of tuition, room, and board) in one's humble one-thousand-square-foot Santa Monica bungalow (which cost

$800,000), call oneself a screenwriter, and feel morally superior to those Iowa State law school grads who work for Monsanto so they can buy five-thousand-square-foot tract mansions in Des Moines.

My point is not that the Hampshire-graduate bungalow owner is hypocritical—we're all hypocrites these days, one way or another—but rather my point is that we cannot track our ever-shifting forms of consumer narcissism if we do not clearly understand how ancient human instincts interact with the modern economy, and how people display their ancient psychological traits through this week's hot new products.

## The Narcissism Premium for Cost-Dense Products

If we do a little price-comparison exercise, we can better appreciate the two faces of fitness-flaunting consumer narcissism. Let's take consumerism at face value as a form of materialism—a way of buying raw matter that has been transformed and patterned for human use. How can we compare prices and value-densities for very different products, ranging from apples to bras to cars to cocaine? We can measure them on two fundamental scales: their retail cost, and the amount of matter they contain. As economics meets physics, we can ask how many dollars per pound a variety of different products cost, and see if any notable patterns arise. The table below gives estimates for a range of products. (The notes on the book's website give more detail about each product's features and how I calculated the cost per pound.)

| Product | $US retail (ca. 2008) per pound net weight |
|---|---:|
| Air | free |
| Tap water (Albuquerque) | 0.0000633 |
| Rice | 0.29 |
| Sugar | 0.34 |
| Gasoline (regular unleaded) | 0.7 |
| Can of soda | 0.8 |
| Apples | 1.6 |
| House (typical suburban) | 2 |

| Product | $US retail (ca. 2008) per pound net weight |
|---|---|
| Television (Sony HDTV) | 6 |
| Car (Toyota Camry LE) | 7 |
| Fitness machine (elliptical) | 7.5 |
| Wine (decent Shiraz) | 9 |
| Pet dog (border collie) | 10 |
| Chair (Levenger) | 11.7 |
| Coffee (Starbucks beans) | 12 |
| Beef (sirloin steak) | 12 |
| Book (hardback) | 12.5 |
| Bicycle (Fuji) | 17 |
| Luxury car (Lexus LS 660) | 20 |
| Blue jeans (Levi's) | 22 |
| Chain saw (Husqvarna) | 37 |
| Human blood | 45 |
| Combat knife (Ka-Bar) | 103 |
| Watch (Timex) | 167 |
| Laptop computer (Dell) | 204 |
| Silver bullion | 225 |
| Telescope (TEC) | 238 |
| Bra (Victoria's Secret) | 240 |
| Handgun (Glock) | 440 |
| Private jet (Learjet) | 460 |
| Music CD | 480 |
| Perfume (Samsara) | 930 |
| iPod Classic (w/o songs) | 980 |
| Fake Columbia U. diploma | 1,090 |
| Cell phone (Motorola) | 1,390 |
| Porn DVD | 1,510 |
| Breast implants | 1,930 |
| Lipstick (MAC) | 2,600 |
| Marijuana | 4,900 |
| $20 bills (currency) | 9,100 |

*Continued*

| Product | $US retail (ca. 2008) per pound net weight |
|---|---|
| Luxury watch (Rolex) | 10,100 |
| Fake diamonds (zirconia) | 13,600 |
| Gold bullion | 14,000 |
| Human kidney (black market) | 16,200 |
| Cocaine | 36,200 |
| Human semen (from donor) | 52,900 |
| Viagra | 53,000 |
| Prozac | 63,000 |
| Heroin | 68,000 |
| Ecstasy | 75,600 |
| iPod Classic (full of songs) | 106,700 |
| Botox injection | 141,600 |
| Real Columbia U. diploma | 1.25 million |
| Real diamonds | 15 million |
| Van Gogh painting | 28 million |
| LSD (pure liquid) | 30 million |
| Human egg (from donor) | 4.5 trillion |

This table reveals some shocking truths. First, there is a rather wide spread of cost densities—an implanted human egg costs about 72 quadrillion times as much per pound as tap water, though the egg is constituted mostly of water, plus some chromosomes, membranes, and organelles. The implanted egg represents genuine evolutionary fitness—successful reproduction itself—the gold standard of human value. It carries the most precious cargo that males desire: high-quality genes from an intelligent, attractive woman. For this, there is little supply and much demand, hence a high price. From the viewpoint of males, these market pressures apply equally whether the egg is obtained from a donor who must be paid by check, or from a wife who must be courted by displaying one's kindness, intelligence, and wealth. For example, the billionaire Ron Perelman's first three ex-wives cost several million dollars per child produced, just in divorce settlements, and not counting courtship and maintenance costs: Faith

Golding (married eighteen years, four children, an estimated $8 million settlement), Claudia Cohen (married nine years, one child, about $80 million), Patricia Duff (married eighteen months, one child, about $30 million). (What it cost these women to bear Perelman and his offspring is harder to quantify.)

Another shocking revelation of the table is how little the basic requirements for survival cost. As Adam Smith observed, the two commodities absolutely necessary for short-term human survival—air and water—are virtually free. Without them, we would die respectively in three minutes or six days. The third commodity necessary for living a few weeks—basic vegetarian food (grains, beans, fruits, vegetables)—is also very cheap, less than $2 per pound. From a survivalist viewpoint, then, everything beyond air, water, and food can be considered a luxury good. Of course, there's much more to evolution than merely survival of the fittest, which is why there are products with higher cost densities.

The basic comforts of modern life—housing, transportation, clothing, basic entertainment—are the next-cheapest goods, costing around a few dollars per pound for a suburban house, Toyota Camry, Levi's blue jeans, or a high-definition Sony television. If you're a carnivore who eats cows ($12 per pound) or a vampire who drinks human blood ($45 per pound), you have to pay a little more for your food. Even fitness in the sense of aerobic endurance costs only $7.5 per pound, for a Vision Fitness X6100 Elliptical Trainer, like the one I use.

We reach the magical realms of consumer narcissism once cost density exceeds that of silver bullion ($225 per pound). Here we find the first products designed mainly for flaunting or faking fitness. We can exaggerate our physical attractiveness (Victoria's Secret bra, breast implants, Guerlain perfume, MAC lipstick), intelligence (TEC telescope, alternative-music CDs), aggressiveness (Glock handgun), or social status (fake Columbia University diploma, cell phone, Learjet).

When cost density approaches that of gold bullion ($14,000 per pound), we see even purer narcissism: luxury status symbols (Rolex watches, iPods that are actually full of iTunes songs, diamonds, real

Columbia University diplomas, van Gogh paintings), luxury drugs for improved appearance and performance (Viagra, Prozac, Botox), and luxury drugs for pleasure (cocaine, heroin, Ecstasy, LSD). Strangely, the patented prescription appearance drugs have about the same cost density as the illegal pleasure drugs. What do these three product classes have in common? They are fundamentally psychological in nature, not material. They all have a fairly direct effect on the owner's brain, or on the brains of observers. They entertain our minds or impress others' minds. They reach right into our nervous systems, grab our attention, jump-start our emotions, and make us exclaim "cool, rad, extreme!" Only a few of these cost-dense products are directly related to survival (kidney transplants) or reproduction (sperm, eggs). Even in the thermospheric heights of cost, well above pure gold's stratosphere, we seem willing to pay as much per pound for fake fitness as for real fitness. (A fake fitness indicator displays inaccurate information to others about one's underlying biological traits, whereas a fake fitness cue conveys inaccurate information to oneself about the object's likely survival or reproductive benefits.)

Note that for many luxury status symbols, much cheaper fake versions exist. Real diamonds have 1,100 times the cost density of high-quality fake diamonds (cubic zirconia, CZ). Real university diplomas have 1,150 times the cost density of fake diplomas. From one viewpoint, the fakes seem hopelessly cheap and tawdry, but from another, the "real" goods seem absurdly overpriced. Arguably, the diamond buyer is overpaying the De Beers cartel by a factor of a thousand compared with the CZ buyer, for an almost indistinguishable stone. Likewise, the Lexus LS 660h L luxury car has three times the cost density of the Toyota Camry—which may imply that the reliable, near-luxury Camry offers three times the value of the Lexus, which is also made by Toyota Motor Corporation.

Of course, cost density is not the only possible way to distinguish luxuries from necessities, or to compare the functions of different products. We could have compared cost per unit of time (hour of enjoyment), taking into account not just goods (such as a Dell laptop cost divided by expected hours of use before obsolescence), but services as well (hourly costs for babysitting, psychotherapy, prostitution, university lectures, or amusement-park rides). Likewise, we could have compared the ratio of final retail price (as for a Lexus car) to the cost of raw material inputs (such as steel, glass, rubber, leather). In either case, we would find much the same pattern: basic survival goods are cheap, whereas narcissistic self-stimulation and social-display products are expensive. Living doesn't cost much, but showing off does.

Product weight is, however, especially misleading for informational goods such as music CDs and porn DVDs, whose contents could be downloaded digitally as very lightweight patterns of electrons. Most such informational goods are also clear cases of self-stimulation or social display, since they can't have any real-world effects except through one's own or other people's senses. For these reasons, they are likely to have extravagant costs that can't be attributed to their material utility.

Finally, many important things are notable for their absence from the cost-density table. Even in the twenty-first century, we still can't buy true love, respect, or fulfillment. If we're lacking them, we can't buy sane parents, successful siblings, or sensible children. We can't even buy decent replacements for biological adaptations that go wrong—artificial eyes, brains, hands, or wombs. Our bodily organs are the most value-dense items that we can call our own. They are beyond price, but we take them for granted until we lose them through accident or age. If you were going blind through macular degeneration, how much would you pay for another ten years of sight? If you were suffocating from emphysema, what would you pay for another one hundred clear breaths? If you were infertile and wanted children, how much would you pay for working sperm or eggs of your own—not just DNA from an unknown donor?

Our inherited legacy of adaptations is literally precious. Even the poorest parents give their children vast riches, in the form of senses, emotions, and mental faculties that have been optimized through millions of years of product development. They are so reliable, efficient, intricate, self-growing, and self-repairing that no technology comes anywhere close to matching them. The human genome is the ancestral vault of riches, the secret Swiss account. It is very important for consumerist capitalism to make us forget this, to take for granted what we owe to life itself. Beyond our true necessities and luxuries—our biological adaptations—we get only a little added value from market-traded products.

Ultimately, the fundamental difference in our existence is not between being rich and poor, but between being alive and not alive, breathing and not breathing. This is why people focus on breathing during meditation: they remind themselves that inhalation and exhalation tower far above Aladdin's palace as gifts for which to be thankful. This is a literal truth that transcends sentimental New Age nonsense; if you move the palace to the airless moon, you won't long enjoy its agate walls and ruby windows. One of my goals is to reveal exactly how evolved human nature engages with our market economy, so we can attach the right relative values to organic adaptations versus artificial products. Fools toast each other's wealth, whereas sages toast each other's health.

## What The Sims 2 Got Wrong About Consumer Narcissism

Consumer behavior is so fundamental to modern life that modern life-simulation games are largely consumption-simulators. We no longer play the 1960 Milton Bradley Game of Life, with its people-pegs in little plastic cars, stopping at the "Get Job" and "Get Married" spaces, trying to reach the road's end (retirement) with the most money. Instead, we play The Sims, released in 2000 by Electronic Arts, which became the most popular computer game in history, selling 6 million copies to date. Its follow-up, The Sims 2, released in 2004, sold more than a million copies in its first ten days, at $50 each.

In The Sims 2, the player controls the behavior of simulated human beings (Sims) who live together in houses that are rendered on-screen in amazing detail. The Sims can be designed from scratch with a wide variety of personalities, facial appearances, and clothing styles. The Sims can get simulated jobs (such as doctor, actor, con artist, or police officer), and make simulated money, using it to buy simulated furniture, appliances, electronics, decorations, and skill-building items (such as books, chessboards, or fitness machines). The Sims can be guided by mouse movements to undertake mundane duties (such as cook, eat, sleep, shower, or go to work) and riskier social tactics (such as tickle neighbors, slap enemies, grope housemates, or greet extraterrestrials). The game is fairly open-ended, with no predetermined goals, no levels to play through, no points to accumulate. Yet there are some implicit success criteria: most players strive to make their Sims happy, wealthy, well employed, and well networked, and to avoid letting their Sims die of hunger, drowning, burning, or electrocution. Technically, The Sims 2 is very sophisticated—a fully 3-D agent-based artificial life simulation that requires more disk space (3.5 gigabytes) than any PC contained in 1995.

Computer-game revenue now exceeds that of Hollywood feature films, and The Sims franchise accounts for a substantial portion of it. The Sims is a major cultural phenomenon. They are the first computer games to crack the gender barrier and successfully appeal to girls and women. They have huge appeal across age groups (from preteens through the retired) and across cultures (being especially popular in North America, Europe, and East Asia). The core games (The Sims, The Sims 2) led to the creation of more successful spin-offs ("expansion packs") than any other computer game—spin-offs that allow, for example, romantic encounters (The Sims: Hot Date, 2001), pets (The Sims: Unleashed, 2002), vacations (The Sims: Vacation, 2002), multiplayer online interaction (The Sims Online, 2002), celebrity status (The Sims: Superstar, 2003), college dorm life (The Sims 2: University, 2005), flirtation in bars and dance clubs (The Sims 2: Nightlife, 2005), running small businesses (The Sims 2: Open for Business, 2006), and enjoying seasonal lifestyle products (The Sims 2: Seasons, 2007).

These games are also potent educational tools, whereby young people can learn a mental model of adult life in ways that formal schooling never offers. To succeed in these games, one must learn how to make and sustain friendships, keep domestic peace among housemates, pay bills promptly, allocate time efficiently, get work promotions through skills and social connections, and manage household renovations and improvements. So far, so good—I'm happy to see my daughter play The Sims 2 and learn how to navigate through the challenges of contemporary life.

Yet the way that human nature is portrayed in these games is worrying. Rather than a single global happiness score, the Sims have eight specific needs to fulfill through buying and using consumer products. Five of them are reasonable survival requirements for a social primate: hunger (satisfied by eating food), energy (satisfied by sleeping), bladder (satisfied by using the toilet), hygiene (satisfied by showering), and social (satisfied by talking in person or on phone). Three of them, though, are a little more nebulous: comfort (satisfied by relaxing, napping, sleeping), fun (satisfied by socializing, playing, or using electronic entertainment), and environment (satisfied by decorating the house with luxury products and art). The whole rationale for studying, working, and buying in The Sims 2 is to fulfill these needs more efficiently, in less time. A more expensive bed boosts energy faster during sleep; a more expensive recliner chair boosts comfort faster; a more expensive shower boosts hygiene faster. Since time is the crucial strategic resource for Sims, product upgrades are all about saving time, rather than flaunting fitness.

Strangely, The Sims 2 does not simulate social status, prestige, or sexual attractiveness, or allow product acquisition to influence these qualities. Buying a more impressive bed, recliner, or shower will not attract more friends or mates. Thus, the social-display aspect of narcissism is completely missing from the game. The subjective-pleasure aspect of narcissism, meanwhile, is hidden behind the three nebulous needs: comfort, fun, and environment. Luxury furniture boosts comfort scores faster; luxury TVs, stereos, and computers boost fun scores faster; luxury paintings, sculptures, lamps, and decorations boost environment scores faster. By presenting these three forms of self-stimulating narcissism as basic human needs on a par with hunger and hygiene, The Sims 2 portrays runaway consumerism as natural human behavior. As in Maslow's hierarchy of needs, the instincts for social display are hidden behind vaguely aspirational drives.

Thus, the most popular computer games in history teach players that bourgeois careerism and endless consumption are the twin pillars of a happy, fulfilling life. The Sims learn, work, and buy, but they do not vote, protest, form unions, do volunteer work, give to charity, or go to church. They are economically empowered but politically neutered. Marx would have viewed The Sims 2 as the most advanced form of cultural superstructure ever developed—3.5 gigabytes of interactive, high-resolution, self-inflicted propaganda supporting capitalist ideology and political apathy. No need for fascist goon squads to make kids play it at gunpoint in public schools; they willingly play it themselves, actually believing it's an escape from the educational indoctrination that they call homework.

Psychologically more realistic life-simulation games would include the two faces of consumer narcissism. Greedy careerist Sims would chase self-stimulation through costly products that promise comfort, fun, and environmental aesthetics, but that do not always deliver. They would chase status cues that promise to advertise their intelligence, kindness, and popularity more effectively to other Sims than ordinary social interaction can. On the other hand, there could also be anti-consumerist Sims, who survive, socialize, mate, and reproduce with basic appliances in small houses, through minimal work, with plenty of leisure. They could enjoy life without buying much, apart from the occasional *Adbusters* magazine or Noam Chomsky video. These Sims could build their own houses and make their own furniture. Instead of buying everything at the full manufacturer's suggested retail price, they could buy things on sale, or used, or from thrift stores. Instead of buying a home fitness machine or swimming pool to exercise, they could run around playing tag, or have tickly sex, for free. Instead of seeking transcendence in front of big-screen HDTVs, they could meditate, pray, or take drugs. They could overturn the atomized suburban lifestyle by joining cohousing estates with communal dining and child care. They could even vote for higher-density mixed-use zoning that combines residential, commercial, and leisure functions in a New Urbanist utopia. Alas, such radical-but-sensible options never appear in the game's vision of life success.

This is not to say that Electronic Arts is part of a global consumer-indoctrination conspiracy. It's just a medium-size company with $3 billion in 2006 revenue, employing seventy-two hundred game developers and support staff who take pride in designing fabulous cutting-edge games. However, those employees are mostly middle-class white American male software engineers, who are living in placid suburbs near Redwood City, California, and who are more interested in employee stock options and the Best Buy megastore than in Buy Nothing Day. Not surprisingly, they tend to channel their own Silicon Valley workaholic-shopaholic values into the needs and aspirations of their Sims. And when we play their games, we tend to internalize those same values so our Sims can prosper.

# 5

# The Fundamental Consumerist Delusion

So, WHAT ARE we really trying to show off with our products? Superficially, consumer narcissism allows people to display their wealth, status, and taste. Yet these are maddeningly vague terms. "Wealth" includes not just assets and income, but the borrowing power to obtain house, car, and business loans. It depends on a little three-digit number called a credit score, which reflects one's history of being a profligate borrower who repays loans conscientiously enough to be dependable, but slowly enough to be profitable to lenders. The major U.S. credit reporting companies (Equifax, Experian, TransUnion) also take into account personal factors such as employment history, residence stability, and debt-to-income ratios. Thus, wealth as borrowing power depends largely on the mental traits of high conscientiousness (which predicts gainful employment, fewer missed payments, and lower bankruptcy risk), and high intelligence (which predicts education, income, and learning enough about the credit system to own the optimal number of credit cards).

Further, not all wealth is seen as morally equal. We make fine discriminations between wealth acquired "legitimately" through the meritocratic, pro-social ideals of individual hard work that helps others, versus wealth acquired through inheritance, marriage, windfall, gambling, or crime. Wealth takes on a different connotation if it is displayed by an organic farmer or brain surgeon, as opposed to an arms dealer, supermodel, lottery winner, Enron executive, Afghan warlord, cult leader, gold digger, or rent boy. We make different attributions about the personality, intelligence, and moral traits of wealthy people based on the sources of their wealth, and pure wealth displays are not generally informative about such traits. To overcome this ambiguity about all the traits beyond wealth, many luxury goods are "positioned" to signal more-specific aspects of the owner's identity (that is, personality traits). For about $50,000, one can buy any of the following new sedans: a new BMW M3, Cadillac CTS-V, Jaguar S-type, Lexus GS 460, or Lincoln Town Car. These five models are designed and branded to convey very different impressions about the owners' traits and their likely sources of wealth, which I shamelessly stereotype as follows:

- BMW M3: divorced 40-year-old male assistant district attorney who needs small rear seats for weekend child custody, and knee-weakening acceleration for dates with leggy criminal defense lawyers
- Cadillac CTS-V: single 19-year-old rap star, recently signed by Interscope Records, soon to lose driving license through DWI convictions
- Jaguar S-type: separated 50-year-old female real estate agent who used to be a pole dancer; proud not to have used OxyContin in the past six weeks
- Lexus GS 460: 35-year-old lesbian professor of cultural studies, recently tenured for a book on the history of condom packaging, cohabiting with female krav maga instructor who drives a Subaru Outback
- Lincoln Town Car: married 75-year-old couple who ran a modestly successful John Deere tractor dealership in Plano, Texas; proud to be American

However, all such positioning requires additional advertising effort to create symbolic associations between the brand and the aspirational traits it embodies, including the specific source and form of wealth of prospective buyers.

"Status" is an even more elusive concept. It basically means anything that provokes social interest, attraction, or deference. In any species of social primate, a higher-status animal is simply one who is looked at and groomed more often by others, who can displace others from desired resources such as food, and who is solicited more often as a friend, ally, or mate. (Robin Dunbar has shown that we humans use verbal grooming—talking—instead of physical grooming to ingratiate ourselves with higher-status individuals.) The question is, what confers status? Certainly products can aim to advertise one's status, and act as status symbols, but they do not actually confer status. That is done by other people: one's status dwells in the minds of observers. Politicians have no more status than what is granted to them by voters, media pundits, and corporate campaign donors. Scientists have no more status than what other scientists award them through citations, talk invitations, and tenure. "Status" makes a misleadingly concrete-sounding noun out of many social verbs distributed among many observers. Status is what we confer on one another—usually through other individuals' judgments on physical, mental, personality, and moral traits. Beauty raises status. Creativity raises status. Emotional stability and articulate leadership during group emergencies raise status.

There are as many types of status as there are types of individual differences between people. Individual differences in intelligence are substantial, stable, and highly predictive of behavioral competence across many domains, so differences in intellectual status exist. Individual differences in kindness and agreeableness are substantial, stable, and highly predictive of altruistic behavior across many domains, so differences in moral status exist. So, here again, when we speak of buying products to display our status, we really mean buying products to display the fact that our physical, mental, or moral traits are superior to those of other people in some comparison group. Like "wealth," "status" boils down to a type of superiority with regard to some set of individual-differences dimensions that have already been noted, judged, and validated by others.

"Taste" admits an even broader diversity of interpretations, as one person's elegance is often another person's kitsch. It is not easy to "flaunt your taste," as the Hennessy cognac ads suggest, in a way that appeals to everyone. This does not mean that taste is all in the eye of the beholder. Rather, it means that taste is a way for us to sort one another out, to choose friends and mates based on similar aesthetic and moral criteria that reflect commonalities of intelligence, personality, and ideology. Common ground in aesthetics, morals, and personality traits make it easier for people to coordinate their behavior with one another for their mutual benefit. Similar tastes make similar stimuli, ideas, and behavioral tactics more salient to each individual. In game-theory terms, they make it easier for people to coordinate on certain "focal points" in "coordination games." For instance, if I arrange to meet an old friend in London on a particular date, but we forget to specify an exact time and place to meet, it would help enormously if we could anticipate one another's tastes and thinking styles. Most people know that meeting at noon is a more salient focal point for finding one another than trying to meet at any other particular time, such as 2:41 a.m. People eat lunch around noon, so meeting at a restaurant might make sense. If my friend and I know we have similar preferences concerning food, price, and location, and know what those preferences are, it's much easier to find each other. We would meet at the Wagamama noodle bar near the British Museum.

Conspicuously displayed aesthetic taste is a convenient, visible way for people to display their deeper personality traits. For example, if I were rich, I would collect paintings by the contemporary artist Fred Tomaselli, rather than the usual Post-Impressionists or Abstract Expressionists collected by Upper East Side hedge-fund managers. Why? Because I find Tomaselli's work visually and intellectually richer, and I appreciate the biological materials, compositional skills, and psychedelic themes. In other words, I would want my art collection to reflect my personal taste, meaning in this case I would (unconsciously) want it to proclaim my openness (to weird hallucinogen-inspired art, and to images of life's spooky transience), conscientiousness (esteem for artists with an obsessive-compulsive attention to detail), and intelligence (appreciation of quasi-conceptual art and knowledge of semi-obscure twenty-first-century artists).

Personal taste should not just attract like-minded individuals; it should also repulse differently minded ones. To be effective, it must be a high-risk, high-gain form of taste signaling, rather than a meek nod to the least common denominator. The Tomaselli paintings would be effective for my social-screening purposes because few people of low openness could bear to sit through a dinner party with such disorienting works on the walls. They would feel existential nausea and never come back. On the other hand, visitors who admired the work articulately, without gagging, would reliably signal their higher openness. Conversely, Christians can repulse atheist intellectuals like me by hanging black-velvet Jesus paintings on their walls, just as Van Helsing repelled vampires with garlic.

Thus, while it is superficially true to say that products display our wealth, status, and taste, these terms do nothing more than dip a timorous toe into the shallows of scientific insight. Real understanding of how we convey our traits through our consumer behavior must, I think, be anchored in a few key facts:

- We are social primates who survive and reproduce largely through attracting practical support from kin, friends, and mates.
- We get that support insofar as others view us as offering desirable traits that fit their needs.
- Over the past few million years, we have evolved many mental and moral capacities to display those desirable traits.
- Over the past few thousand years, we have learned that these desirable traits can also be displayed through buying and displaying various goods and services in market economies.

The most desirable traits are not wealth, status, and taste—these are just vague pseudo-traits that are achieved and displayed in widely different ways across different cultures, and ones that do not show very high stability within individual lives, or very high heritability across generations. They exist at the wrong level of description to be scientifically useful in connecting consumer psychology to evolutionary psychology. Rather, the most desirable traits are universal, stable, heritable traits closely related to biological fitness—traits like physical attractiveness, physical health, mental health, intelligence, and personality. When we really want to find out about someone—as a potential friend, mate, co-worker, mentor, or political leader—these are the traits we are most motivated to assess accurately. Consumerism's dirty little secret is that we do a rather good job of assessing such traits through ordinary human conversation, such that the trait-displaying goods and services we work so hard to buy are largely redundant, and sometimes counterproductive. This raises the question: Why do we waste so much time, energy, and money on consumerist trait displays?

## The Social Psychology of Consumer Narcissism

The whole edifice of consumer narcissism rests on the questionable premise that other people actually notice and care about the products that we buy and display. Sometimes they do, but often they don't, and we overestimate how much they actually do. This is a deep failure of human social psychology. Under natural conditions, we are generally rather good at doing perspective-taking—imagining other people's points of view, and understanding what they notice and care about. Although we humans are better at perspective-taking than any other animal, we are far from perfect, especially since we have come to live under evolutionarily novel, unnatural conditions, such as being swamped in consumerism and spoiled for choice amid branded products.

Advertisements for most products converge on one key message: other people will care deeply what products we buy, display, and use. At first glance, this message sounds absurd—socially implausible and easily disproved by talking with others. However, given that we're exposed to about three thousand ads per day repeating some version of this message, it's hard to remain skeptical. The result is that we greatly overestimate how much attention others pay to our product displays, through which we are unconsciously striving to show off our key bodily and mental traits. We also underestimate how much attention others pay to more natural forms of trait display that can be judged easily and accurately in a few minutes of observation and conversation.

Seriously, can you remember anything specific worn by your spouse or best friend the day before yesterday? Can you remember what kind of watch your boss wears? The brand of your nearest neighbor's dining room table? The face of the last person you saw driving a Ferrari? Probably not, unless you have the obsessive consumer fetishism of *American Psycho*'s protagonist. Mostly, we just don't care what kinds of products strangers display, except for a few domains in which we have a professional or personal interest: dentists notice your teeth; hobbyist jewelry makers notice your earrings. (We may know what kinds of cars and clothes our friends and mates own, but we've already learned about their deeper traits anyway, so their product choices don't carry much further information.)

In fact, decades of social psychology research suggest that we automatically notice only a few basic traits when we see people: their size, shape, age, sex, race, familiarity, relatedness, and attractiveness. We also notice special states of physiology (sleep, injury, sickness, pregnancy) and emotion (anger, fear, disgust, sadness, elation). Throughout human evolution, these have been the most significant things to notice about people, because they carry the most crucial implications about how we should interact with them to promote our own survival and reproductive prospects. It was always important for females to distinguish between their babies, sisters, boyfriends, and stalkers. It was not so important that they notice exactly what kind of furs, beads, or body paint each individual was wearing, except when they were useful in assessing a stranger's social status.

Indeed, the traits that are most salient and relevant to people are precisely the traits that remain hardest for purchased products to signal reliably—or to misrepresent credibly. It is very difficult to buy goods or services that can notably alter one's apparent age, sex, or race, or that can disguise one's broken leg, oral herpes, or basic emotions. A $15,000 face-lift can make a fifity-five-year-old woman look more like a thirty-five-year-old with regard to facial sagging and wrinkles, but cannot hide other cues of age on the neck and hands. A sex change through hormone therapy and sex reassignment surgery (also about $15,000) can transform apparent sex in some ways, but has little effect on height, torso shape, facial bone structure, or the sexually differentiated brain that grew in utero—much less on one's ability to reproduce as a member of the opposite sex. In each of these situations, our social-perceptual systems for recognizing key human traits and emotions are hard to mislead, because they have been evolving so long to be accurate. They have become very efficient at vacuuming up all the information they can from all the different cues that can be perceived from an individual's body, face, language, and behavior.

After we notice people's key demographic and physical traits, we seek information about their mental traits. We want to know a few basic things about how their brains work. How intelligent and mentally healthy are they? What kind of personality do they have? What moral virtues do they signal through the political and religious beliefs that they espouse? These, again, are the traits that carry the most predictive information about how to interact with someone. They are also the traits that we have evolved to assess most quickly and reliably through the normal prehistoric modes of interaction: greeting, eating together, and talking. And again, they are the traits that are hardest to fake through bought products—though the rest of this book basically details how we try to do so, in various self-deluded and ineffectual ways.

Recent research on "person perception" suggests that we are really rather good at judging other people's intelligence, sanity, and personality from just a few minutes of observing their behavior or talking with them. Accuracy can be measured by determining how consistent different personality cues are with one another ("convergent validity"), the degree to which people agree when judging a particular trait ("inter-rater reliability"), and how well the trait judgments predict an individual's future behavior ("predictive validity"). Accuracy tends to be higher for more visible traits such as extraversion (talkativeness and outgoingness), and lower for more internal traits such as neuroticism (tendency to worry, ruminate, and feel anxious). Accuracy is also higher when we judge a person behaving in a free, unscripted situation that allows individual differences to reveal themselves (as when chatting at a party or living in a small-scale hunter-gatherer group) than in a situation highly structured by social norms that suppress individual differences (as when standing in line at a cash machine or marching in a military parade). Observed behavior also carries more reliable information when the persons in question believe they are alone, and are not constructing a false persona for public approval. This is why the sight of men helping the elderly or rescuing kittens is so attractive to women, especially when the men do not know they are being watched.

The personality psychologist David Funder has summarized many of these effects in his realistic accuracy model of person perception. His model posits that there are plenty of more-or-less objective, reliable behavioral cues available to inform us about most personality traits. Our accuracy at judging those traits depends simply on others' doing or saying things that express these relevant and informative cues, and on our noticing, perceiving, and judging the cues appropriately. These personality cues can include every aspect of how people talk, move, and dress—everything from how they pronounce "Goethe" to how they discuss Neil Gaiman's graphic novels; everything from their walking speed to their erotic eloquence when dancing a tango nuevo.

Funder's model also implies that when personality-relevant information is not readily available about someone we wish to judge, we may often create social occasions in which the information becomes more available. For example, it may be hard to judge a daughter's boyfriend's agreeableness (kindness, warmth, generosity) if we meet him in a quiet, air-conditioned steak house. Much better to invite him over for a midsummer extended-family barbecue at which he is in encouraged to drink several beers, and then assaulted chaotically on all sides by children, dogs, footballs, and stinging insects. If, under these more difficult, disinhibited, and diagnostic conditions, he becomes irritable to the point of throwing the footballs at the dogs and squirting mustard at the children, we know his agreeableness level is rather low (and that he might have a short temper with our daughter's future babies). Conversely, if he remains calm, cheerful, and helpful as the sweat rolls down his beer-flushed, mosquito-stung, dog-licked face, we know his agreeableness level is rather high. The cultural evolution of such occasions for accurate personality assessment may explain why major social rituals (dates, job interviews, parties, banquets, holidays, weddings, honeymoons) entail such long durations, high stress levels, and disinhibiting drugs such as alcohol. These conditions bring out both the best and the worst in us.

Many mental disorders are also rather easy to detect within a few minutes, on the basis of appearance, behavior, and conversation. People with major depression tend to slump and look sad; they speak softly, slowly, and monotonously; they disparage their lives and prospects. People with schizophrenia tend to be unwashed, ungroomed, and dressed in too many layers of clothing; they have odd gaits and mannerisms; their speech is sometimes incoherent, rambling, and delusional. Other easily observed cues characterize people with anxiety disorders, obsessive-compulsive disorder, autism, anorexia, narcolepsy, and most personality disorders. Only a few mental disorders are really hard to identify from superficial interaction: psychopathy, specific phobias, sexual disorders and dysfunctions, and some addictions. When it comes to judging people's sanity, most experienced adults are rather accurate. We may not be able to diagnose each peculiarity using the current psychiatric terms, but the basic difference between normal and abnormal behavior is highly salient. This is especially true when we are making judgments about people who may fill central roles in our lives: potential mates, friends, business partners, and in-laws. Even psychopathy—the disorder that is hardest to detect through short-term individual interaction—can sometimes be detected through gossip about the psychopath's previous crimes and misdemeanors.

## The Fetishization of Youth and Disparagement of Wisdom in Consumerist Social Judgment

The accuracy of person perception tends to improve with age, as we learn, gradually and painfully, which behavioral cues are the most reliable indicators of personality, intelligence, and moral virtues. We learn which situations reveal the most diagnostic information about someone's true character. We learn how to see through first impressions.

This explains why the dating choices made by teenagers have always seemed appallingly stupid to their parents. Teenagers are overly influenced by the traits that are easiest to assess (physical attractiveness and status among peers). By contrast, parents have decades more experience in assessing the harder-to-discern traits, such as conscientiousness, agreeableness, emotional stability, and intelligence, and in appreciating the longer-term benefits that these traits convey in any human relationship. This ability to judge character was considered a major part of wisdom, and a cardinal virtue, before consumerist capitalism made concepts like character, wisdom, and virtue sound unfashionable.

Why was evolution so remiss in failing to arm human teenagers with sensible mate-choice preferences? One answer is that their preferences may be rather well-adapted to getting good genes in the context of short-term mating, even if they're not so good at finding good partners for the sort of long-term relationships that yield higher social and economic benefits under modern conditions. A second answer is that teenagers reach puberty far earlier today (probably due to higher-fat diets) than they did under prehistoric conditions, when their sexual psychology had more time to catch up to their sexual physiology. A third answer might be that parents always had a fairly heavy influence on mate choices made by their teenage offspring, so evolution focused on shaping the parents' preferences rather than the teens'.

In any case, by the mid-twentieth century, it became crucial for marketers to convince young people that they could judge one another's individuality more effectively through consumerist trait displays than their elders could through wise observation. Judgments of one's peers and dates by the older generation had to be made to seem old-fashioned, uncool, irrelevant, biased, and prejudiced. In this, the marketers succeeded spectacularly, assisted by two key twentieth-century ideologies: (1) the egalitarian rejection of the idea that an individual's personality, intelligence, mental health, and moral virtues are useful concepts worth evaluating accurately and discussing socially, and (2) the environmentalist rejection of the idea that these traits show stability within individuals (across situations, relationships, and ages) and within families (through genetic inheritance).

Consumerist capitalism has depended on youth's embrace of these blank-slate ideologies, which were sold as thrillingly rebellious and thoughtfully progressive. Throughout most of the twentieth century, they seemed validated by psychology, social science, progressive politics, and the self-help movement. In popular culture, the blank-slate ideology convinced the young that the purchase of any new product designed to display some personal trait was a heroic rebellion against the older generation's outmoded belief in the existence, stability, and heritability of personal traits. In the behavioral sciences, the blank-slate ideology biased generations of scientists against trait psychology, personality research, intelligence research, behavior genetics, and any other area concerned with individual differences. Instead, the focus turned to psychological processes that were allegedly similar across all humans: child development, social cognition, neural information processing.

As long as advertising never actually used the old-fashioned terms for traits (character, intelligence, virtue), the young could buy, display, and admire the trait-displaying products, make the social judgments they needed to make about one another's traits, and pretend that they were living in a radical new post-trait world. The whole discourse of traits went underground, discreetly hidden in the rhetoric and semiotics of branding and marketing. It remained just visible enough for the young to recognize, unconsciously, which products would display which traits, but it was just elusive enough that their anti-trait ideology was never threatened, and the person-perception wisdom of their parents never seemed relevant to their lives. For example, rap music producers such as Dr. Dre realized in the 1990s that the real money lay in convincing white middle-class suburban boys that by buying and playing rap, they could display their coolness, attitude, and street cred (that is, their aspirations toward low conscientiousness, low agreeableness, and high promiscuity). The white boys obliged by pouring billions of their parents' dollars through the local music retailers' hip-hop sections, while dissing their parents' concerns that white girls might actually prefer to date boys who display high conscientiousness, agreeableness, and chastity. But if the parents couldn't distinguish between DJ Spooky, DJ Spinna, and DJ Qualls, how could they possibly claim that the whole rap music industry was just another marketing-driven set of costly, unreliable trait displays, or that the trait displays their children considered cool were actually repulsive to potential mates, friends, and employers?

Thus, the blank-slate model of human nature, far from challenging the principles of consumerist capitalism, forms consumerism's ideological bedrock. It makes the trait-perception wisdom of older generations seem outdated and irrelevant, and makes the trait-display aspirations of younger generations seem to require buying the appropriate goods and services, while allowing them to pretend that they live in a brave new post-trait world. Most importantly, it undermines everyone's confidence that their traits are real enough and visible enough to be appreciated without being amplified and externalized by careerism and consumerism.

## The Fundamental Consumerist Delusion

Consumerism depends on forgetting a truth and believing a falsehood. The truth that must be forgotten is that we humans have already spent millions of years evolving awesomely effective ways to display our mental and moral traits to one another through natural social behaviors such as language, art, music, generosity, creativity, and ideology. We can all do so without credentials, careers, credit ratings, or crateloads of product. Our finest, most impressive goods and services have been endowed to us by our DNA, in the form of physical and psychological adaptations that naturally display our virtues and naturally impress our peers. Our ancestors unwittingly invested enormous effort into building up this genetic legacy of accurate trait display through billions of attempts over millions of years to win friends, influence people, court mates, choose mates, raise children, and show magnanimity to one another. This is a core message from evolutionary psychology: the most precious, complex, intricate, and wonderful things in life are the biological adaptations common across all humans—especially the adaptations that signal our individual differences so conspicuously. We already have everything we could possibly need to impress our fellow humans, yet every major human ideology conspires to make us forget this fact—because every ideology seeks power by convincing us that we need something beyond our naked bodies and minds to be socially acceptable and sexually attractive. Consumerism has become our most potent ideology because it so contemptuously dismisses our natural human modes of trait display, and it keeps us too busy—working, shopping, and product displaying—to remember what we can signal without all the products.

Consumerism actually promotes two big lies. One is that above-average products can compensate for below-average traits when one is trying to build serious long-term relationships with mates, friends, or family. True, some products can mask personal defects in the short term. For a forty-seven-year-old single woman seeking mates, Botox can paralyze facial muscles to reduce wrinkles, hiding some signs of age. The treatment might lead a thirty-one-year-old single man to ask her for a second date, which he might not have done had her true age been more apparent during a candlelit first date. However, that age will become apparent sooner or later through other, more reliable cues: the appearance of hands and neck in daylight, an introduction to her twenty-five-year-old daughter or fifty-two-year-old sister, the invitation to her thirtieth high school reunion, and so on. The same principle applies to almost every other product that tries to enhance physical appearance, apparent intelligence, personality, or moral virtues. Trait-enhancing products can fool some of the people in the short term, but they can't fool any of the people in the long term. This is why newlyweds are more often disappointed than delighted to discover their spouse's true character during stressful foreign honeymoons.

A second big lie that consumerism promotes is that products offer cooler, more impressive ways to display our desirable traits than any natural behavior could provide. Specifically, consumerism assumes that better products are more effective signals. Any technical improvement in product design or features, and any marketing innovation in product branding, is pitched as an upgrade in signal effectiveness. Indeed, if we buy products primarily as signals of our underlying biological traits, their signal effectiveness—especially as carried by brand recognition—is, logically, paramount, while their efficiency in serving their nominal purpose (as a garment, appliance, or vehicle) is only of secondary concern. This fact is perfectly clear to every marketing professional, but it must remain perfectly obscure to most consumers. Advertising must therefore play a coy and subtle game with the consumer: while it must hint at the signaling functions of conspicuous consumption, it must never make quantitative claims about the relative signaling efficiency of different products, or of artificial products versus natural human behaviors.

Such explicit claims about a product's trait-signaling power could be proven false all too easily. For example, sports car ads aimed at single males must imply that driving the car will result in the males' attracting more attention from beautiful young women. But the ads must not make that claim explicitly, because it would be too easy for advertising regulators or rival manufacturers to demonstrate empirically that the sports car's drivers do not enjoy a sufficient increase in attention to justify its price premium, and that a better sense of humor would increase female attention far more effectively than excess horsepower.

Thus, consumerist capitalism must keep the signaling functions of products at the stage's periphery, in the shadowy netherworld of smoke, mirrors, curtains, veils, video girls, and dream boys. Direct claims that a product will increase one's social popularity or sexual attractiveness could not withstand the spotlight of center stage, where the quality of one's performance can be judged all too harshly. Instead, the product's nominal functions, features, specifications, novelties, popularity, and branding must occupy the consumer's conscious attention, while the promise of signaling status and sex appeal must penetrate the unconscious as silently and unaccountably as a stealth bomber. Consumers must feel that they uniquely recognize the signaling potential of the product from the subtext of the ad, that their desire for social status and sex appeal is subjectively legitimate but publicly embarrassing, and that they alone can convert the product's technical excellence into a display of personal coolness that yields social and sexual payoffs. The consumers must feel that they can enter into a signaling conspiracy between themselves, the product, and some hypothetical audience of admirers, and that this conspiracy is racy, transgressive, ingenious, and somehow even subversive of capitalism itself.

Even the consumer rights, education, and protection movements conspire to promote this delusion. While *Consumer Reports* magazine takes great pains to assess empirically the objective features, functions, safety, and reliability of products, it never assesses their signaling efficiencies in promoting the consumer's social reputability or sexual success. Given progress in social science and consumer-research methods, that would actually be rather easy to do, through interviews, questionnaires, and focus groups, in which people rate how they would react to specific individuals' buying, using, and displaying the various products, compared with other competing products, and compared with other possible behaviors.

For example, *Consumer Reports* readers could complete annual questionnaires asking not just how often their Corvette Z06 has needed brake repairs, but whether the Corvette has actually resulted in any new friendships, business partnerships, dinner party invitations from neighbors, or spontaneous sexual encounters with admiring female pedestrians. Even if male Corvette drivers do manage to attract a little extra female attention, the math doesn't work out very well for them. Suppose a male driver enjoys an average of one extra short-term mating per year attributable to his choice of car. The Chevrolet Corvette Z06 ($70,000) has a $50,000 price premium over the comparable-size Chevrolet Malibu sedan ($20,000), and both cars are designed to become obsolete in about five years. Rational car-buyers could then calculate that the Corvette's price premium of $50,000 yields an expected five extra sexual encounters during its five-year product life, or $10,000 per encounter. By contrast, a typical encounter with a professional sex worker costs about $200, or fifty times less. Instead of paying the Corvette's price premium, which might yield one encounter per year, the driver could just buy the Malibu and, with the cash he saved, have one encounter per week. The prospective male Corvette-buyer must accordingly either be wildly overoptimistic about the car's attractiveness to women, or be very bad at math, or strongly prefer sexual encounters with amateurs rather than professionals.

Alternatively, the Corvette coveter may be a husband seeking plausible deniability regarding the car's fantasized role in extramarital sexual adventures—a situation that is probably all too common. Since most consumers spend most of their lives married, the only way to sell products that promise increased sex appeal is to make such pitches below the radar of spousal jealousy. Thus, the *Sports Illustrated* ad for the Corvette must not say, "This will increase your short-term copulation opportunities" (or "This will get you laid"), but it can list some technical specs and show a female passenger throwing up both hands in ecstatic surrender to the 505-horsepower engine and its master. Gullible wives will worry less, and gullible husbands will fantasize more.

Similar subtexts appear in advertising aimed at female consumers. A recent *Vogue* ad for a L'Oréal lipstick called Glam Shine Dazzling Plumping Lipcolour touted its "unique micro-crystal technology" and claimed that its "moisture-drenched formula with non-sticky texture delivers full, healthy lips with dazzling dimension and incredible shine." This breathless techno-sensualism could be rendered more honestly as: "This lipstick will signal your libidinous desperation and imminent ovulation not only to your sexually jaded husband, but to your male neighbors and household servants." Because such direct language might alarm the casual *Vogue*-reading husband and teenage children, however, for the sake of marital and family harmony, "tasteful" ads in mainstream media obligingly conceal the diffident status seeking and sexual caterwauling behind most product pitches. Married consumers can thereby delude one another that they are buying premium products for their utilitarian performance (505 hp or microcrystal technology) rather than their signaling power (to attract thrill-seeking neighbors or sweetly pliable pool boys). As usual, plausible deniability and adaptive self-deception allow human social life to zip along like a maglev monorail above the ravines and crevasses of tactical selfishness, by allowing the most important things to go unsaid—but not unimagined.

From my perspective as an evolutionary psychologist, this is how consumerist capitalism really works: it makes us forget our natural adaptations for showing off desirable fitness-related traits. It deludes us into thinking that artificial products work much better than they really do for showing off these traits. It confuses us about the traits we are trying to display by harping on vague terms at the wrong levels of description (wealth, status, taste), and by obfuscating the most stable, heritable, and predictive traits discovered by individual differences research. It hints coyly at the possible status and sexual payoffs for buying and displaying premium products, but refuses to make such claims explicit, lest consumer watchdogs find those claims empirically false, and lest significant others get upset by the personal motives they reveal. The net result could be called the fundamental consumerist delusion—that other people care more about the artificial products you display through consumerist spending than about the natural traits you display through normal conversation, cooperation, and cuddling.

That this view is delusional should be obvious to any adult who has enjoyed long-term relationships with relatives, friends, lovers, partners, collaborators, mentors, and students. Long-term relationships grow and endure through complex, ever-shifting sets of partly conflicting, partly overlapping interests. They are what economists call repeated-interaction mixed-motive games. They typically include repeated cycles of cooperation and conflict, trust and betrayal, intimacy and alienation. These cycles are rarely influenced by product purchases, but rather by the kinds of arguments, explanations, apologies, resolutions, and gossip that have driven human social life ever since language evolved a few hundred thousand years ago. In the first few minutes of meeting someone, all the stage sets, props, and costumes of consumption may seem salient, and the ballads and dirges of that three-voiced chorus—wealth, status, taste—may seem momentous. Yet after relationships are established, these fade into the background of consciousness, and our attention centers rightly on the characters, actions, words, and relationships that really matter. We recognize our roles as mutual protagonists striding life-size across one another's sets and stages. With friends and family, we ride the zeppelins of joy and the barges of duty. We become keenly alive to the social dimension of human existence. In that naturally social state of mind—a state plausibly typical of our ancestors' every waking hour—the fundamental consumerist delusion that products and brands matter, that they constitute a reasonable set of life aspirations, seems autistic, infantile, inhuman, and existentially toxic.

# 6

# Flaunting Fitness

SINCE ABOUT 1990, there have been two bloodless but momentous revolutions in human affairs: the collapse of communism in politics, and the rise of signaling theory in biology. Both depended on the same insight: individuals work hard mostly because they want to show off to others, not for the good of the group. This tendency holds true in both organic evolution and human economics. Bowerbirds build elaborate nests to attract mates for themselves, not to improve the public aesthetics of their New Guinea habitat. Likewise, Ukrainian farmers will work harder to buy status symbols for themselves than they will to feed starving neighbors.

The basic insight of signaling theory is that animals make a lot of noise about themselves, but they don't communicate much news about the world. We've known ever since Darwin that animals are basically machines for survival and reproduction; now we also know that animals achieve much of their survival and reproductive success through self-advertisement, self-marketing, and self-promotion. Narcissism is nothing new; it is the evolutionary norm, as every peacock strives to be the brand most favored by peahens.

Almost all animal signals—birdsong, firefly lights, pheromones, courtship dances—convey self-promoting information about the signaler, not helpful information about the environment. Instead, most animal signals convey little more than the individual's type (species, sex, age) and quality (fitness, health, status, fertility). They do not include the sort of detailed product specifications favored by 1950s car ads. They just say what kind of beastie you are, and how good you are at being your sort of beastie. Animals send such signals to just a few different audiences for just a few reasons—mainly to solicit care or food from parents, to threaten rivals, and to attract mates. This is accomplished not by conversing about topics of mutual interest, but by sending credible claims about one's own needs ("Gimme food!") or qualities ("I've got great genes, so you should mate with me").

It's easy to claim that one is helplessly hungry or awesomely fit; the challenge is to make such claims credibly. This is precisely what signaling theory addresses: how animals can back up their claims, how they can send reliable, hard-to-fake signals that will be believed. In 1975 the Israeli biologist Amotz Zahavi proposed that high cost could guarantee the reliability of quality signals. His "handicap principle" suggested that only high-quality animals could afford to waste a lot of time, energy, and resources on issuing costly signals, which he called handicaps. A sickly, starving, parasite-ridden, brain-damaged pigeon can't repeat the "mate with me" song a few thousand times an hour; therefore, any pigeon who can repeat the song must not be sickly, starving, or otherwise impaired. The song's cost guarantees the singer's quality. This theory remained controversial until around 1990, when biologists understood it clearly enough to develop mathematical models demonstrating that it works. Since then, Zahavi's handicap principle has expanded into modern "costly signaling theory," which is a foundation of modern research on animal communication, sexual selection, social interaction, and human behavior.

When animals use physical traits or behaviors to show off, we can call these handicaps, or costly signals, or sexual ornaments, or fitness indicators, the term I prefer. As we saw in chapter 5, fitness indicators function as both advertisements and warranties: they not only proclaim quality, but guarantee it. Indicators attract attention if they are costly, hard to produce, and hard to fake. They are ignored if they are too cheap, simple, and easy to counterfeit.

The peacock's tail is the classic example of a fitness indicator. It has no survival function, and plays no necessary role in fertilization. It simply attracts peahens by showing off the peacock's health and fitness, the quality of its genes, and its ability to find seeds and insects, and to escape from tigers. Other obvious fitness indicators include the lion's mane, the elk's antlers, and the humpback whale's song. Human bodies are also full of fitness indicators that reveal reliable information about health and fertility, and that were shaped in part by sexual selection to attract mates. These bodily signals of quality include our faces, voices, hair, skin, gait, and height—plus female breasts, buttocks, and waists, and male beards, penises, and upper-body muscle mass. Many human mental traits may have also evolved as fitness indicators, including our capacities for language, humor, art, music, creativity, intelligence, and kindness.

Signaling theory applies equally to nature and culture. Nature produced peacock tails: large, symmetrical, colorful, costly, awkward, high-maintenance, hard-to-fake fitness indicators. Human culture produces luxury goods like the Hummer H1, which is also large, symmetrical, colorful, costly, awkward, and high-maintenance. These qualities likewise make it hard to fake as a wealth indicator—even if you could steal an H1, you probably couldn't afford its gas or insurance.

## Counterfeiting

Signaling theory becomes clearer when you think about counterfeiting money. Counterfeiting became much easier in the 1990s with digital color copiers, printers, scanners, and computer graphic software—technology that challenged the U.S. Bureau of Printing and Engraving to include stronger anticounterfeiting features in the next-generation redesign of U.S. currency. The $20 bill is an especially popular counterfeiting target, since it's the highest-denomination bill commonly accepted by cashiers without close inspection. Thus, the upgraded Series 2004 $20 bill included several new security features such as a broader range of ink colors, watermarks and microprinting that are very hard to copy or scan, color-shifting ink that is hard to imitate, security threads with microprinting that glow different colors under UV light, and highly detailed, enlarged, off-center portraits that are easy to recognize but hard to imitate.

The common denominators in these security devices are conspicuous cost and conspicuous precision. The bureau's high-speed, sheet-fed rotary Intaglio printing presses are just too expensive for the typical counterfeiter—although once the bureau is equipped, the marginal (unit) cost of printing each $20 bill is only four cents. As for conspicuous precision, the microprinting, security threads, watermarks, and detailed portraits are very complex, detailed, and hard to imitate accurately. When the European Central Bank released 14.5 billion new bills of euro currency on January 1, 2002, they included similar anticounterfeiting features, plus even harder-to-fake iridescent ink and a hologram foil stripe or patch.

To understand the costly signaling theory that explains much of consumption, it's easiest to consider how we distinguish "real" products from "fake" products—and why we care about the difference. Conspicuous cost and conspicuous precision are the two basic features of hard-to-fake signals. Zahavi's handicap principle focused on production cost, and the money printers focus on production precision, but most credible signals—and luxury goods—include high levels of both, and it is both that cheap imitators try to fake.

Consider gold necklaces, for example: their value depends on the weight and purity of their gold and the quality of their workmanship. Gold-plated and hollow-gold necklaces seem fake compared with solid-gold ones simply because they contain fewer atoms of gold. Likewise, 10k necklaces (41.7 percent gold—the minimum allowed to be labeled "real gold" in the United States) seem fake compared with 18k necklaces (75 percent—the minimum for France and Italy). Also, necklaces cast with pits, bubbles, rough edges, uneven color, and poor soldering seem fake compared with well-made necklaces. In each case, the fakery can be detected easily by professionals. Gold content can be assessed by the magnet test (gold isn't magnetic; many base metals such as iron are), the heaviness test (gold is dense, about twice as heavy as base metals), and the acid test (gold does not react to pure nitric acid, but will react to aqua regia—a mixture of nitric and hydrochloric acid). Casting flaws are easily discernible through a 10x triplet loupe. For a cheap thrill, go to your local mall's jewelry chain store with a magnet, a

$15 loupe, and $5 bottles of nitric acid and aqua regia, and ask to test their alleged 14k gold jewelry. This is just the sort of testing we do when trying to discern whether a potential mate or friend is, metaphorically, pure gold or merely gold-plated. We don't use magnets and loupes, but we unconsciously draw upon even more potent means to assess their quality: our evolved capacities for conversation, face perception, and personality inference.

As another example, consider fake Rolex watches. Computer-controlled machine tools, Swiss movements, and cheap sapphire crystals have enabled small-scale East Asian manufacturers to flood the online market with ever better, cheaper imitations of luxury-brand watches. This leads to a signaling arms race: Rolex adds more and more anticounterfeiting features, and the Rolex-imitators learn better and better ways to replicate them. For example, a high-quality $1,200 replica of the Rolex President watch from Replicagod.com is rather hard to distinguish from the original (which costs about $30,000), because both include a waterproof, shock-resistant Swiss ETA 25-jewel movement, a micro-laser-etched crown on the dial, a quad-wrapped 18k gold forged case, a scratchproof sapphire crystal, a 2.5x date magnifying viewer, unique serial and model numbers between the lugs, Luminox hour markers, a black Triplock O-ring seal on the winding crown tube, and a Rolex brand hologram sticker. Unless you read the expert Rolex identifying guides by Richard Brown or John Brozek, it's very hard to tell a real from a fake Rolex—or indeed to justify spending the extra $28,800 for a real one. Similar replica problems afflict all the other luxury watch companies, including Breitling, IWC, Omega, Patek Phillipe, and TAG Heuer. (Surprisingly, a similar problem is emerging in the auto industry: Shuanghuan Automobile in China has been able to produce and sell cheaper vehicles that bear very close similarities to the Honda CR-V, the SmartCar, and the BMW X5.) Likewise, over the course of human evolution, as our capacities for judging others have improved, our capacities for deceiving others have improved in turn, in a never-ending arms race of social judgment and social pretense.

The arms race between real and fake has also undercut the De Beers diamond cartel for more than a century, as ever-better imitation diamonds have been developed: titanium dioxide (synthetic rutile) in the 1940s, synthetic strontium titanate (Fabulite) in the 1950s, yttrium aluminum garnet (YAG) in the 1960s, gadolinium gallium garnet (GGG) in the 1970s, cubic zirconia (CZ) in the 1980s, and silicon carbide (Moissanite) in the 1990s. CZ makes an excellent imitation diamond; I've made a necklace for my daughter from a perfect, three-carat, brilliant-cut CZ bought for $4 from a local gem and mineral shop. Moissanite, introduced in 1998, is even closer to diamond, with a similar hardness, density, and luster, yet more brilliance (a higher refraction index) and more "fire" (a higher dispersion index). Moissanite's manufacturer, Charles & Colvard, advertise it as "not a diamond substitute" but as "an entirely new option in affordable luxury," which "offers the value, quality, and fashion that self-purchasing women demand, without the emotional heft of diamond." That is, women can buy themselves Moissanite rings without having to deal with engagement to "emotionally hefty" men. Casual observers can't tell them apart, nor can most pawnshop owners using the standard thermal conductivity tests for distinguishing CZ from diamond. Only experts may notice the subtle double refractions (birefringence) caused by Moissanite's hexagonal crystal structure. Synthetic corundum is even more annoying to jewelers, since it is exactly the same aluminum oxide as real rubies and sapphires, but it can be made larger, purer (free of "inclusions," also known as dirt), and more evenly colored. Thus, "real" rubies are inferior to synthetic rubies by any rational measure, though they cost a thousand times as much, because their rarity makes them more desirable to some. These advances in gem production raise the possibility that in biological evolution, too, traits that began as fake alternatives to certain signals of quality may have evolved to be more useful and even more desirable than the original traits ever were. For example, verbal humor may have originated as a way for subordinate youths to imitate and mock older, more physically dominant sexual rivals—until eventually, humor became even more attractive than dominance, just as Moissanite achieved higher brilliance and fire than diamonds.

Finally, consider the problem of provenance in art history. Rembrandt painted about 700 pictures, of which only 3,000 are still in existence. . . . Uncannily, Rembrandts continue to proliferate like rabbits, long after his death. Suppose you buy an alleged Rembrandt original for a few million dollars, and very much enjoy showing it off at dinner parties, and discussing its subtleties of form and shade. Then your insurance company's experts determine that it was produced by a talented nineteenth-century forger. The painting has not changed as a physical object—its subtleties of form and shade should remain equally laudable—but its value may have dropped a thousandfold. Why? Because it is no longer "real," no longer in the small category of Canvases Actually Painted by Rembrandt van Rijn, born 1606. Value depends on supply and demand; there is a much larger supply of fake than real paintings by Dutch Old Masters, and a much smaller demand. (Provenance also matters in mate choice: given two possible spouses of equal apparent quality, we generally prefer the one who comes from a higher-quality family, meaning a family full of more successful and desirable blood relatives. Those relatives carry some of the same genes as the potential mate, so we assess them unconsciously as a genetic guarantee of the mate's true quality.)

For every kind of high-value product—paper money, gold necklaces, luxury watches, diamonds, Rembrandt paintings—there is an endless struggle between the real and the fake, the genuinely valuable and the counterfeit. The real products tend to include ever more conspicuous costs (in raw materials, equipment, time, energy, and innovation) and ever more conspicuous precision (symmetry, regularity, complexity, fit, and finish). In response, manufacturers of fake products find ways to substitute cheaper materials and equipment, to minimize production time and energy costs, and to emulate the precision and branding of high-quality goods. The fake ultimately illuminates and challenges the real, as consumers begin to question why they should pay the "real" product's premium. Why bother with a real $8,000 3-carat diamond for an engagement ring, when a $4 CZ stone is indistinguishable to most people? Why bother with a real Rembrandt for $10 million when you can download a high-resolution digital image of one and commission your local FedEx Office store to make a visually indistinguishable full-size giclée print of it (with computer-printed color ink on real canvas) for about $200? The fakes reveal what a high proportion of the real products cost: a luxury brand markup, a pure profit premium, a con. The irony is that, with regard to purely pragmatic value, the "real" version of the product is a bigger rip-off than the "fake" version.

Signaling theory is best understood by thinking about these issues of counterfeiting, cost, precision, and luxury branding. If you look at any costly signal, any fitness indicator, you're always looking at a snapshot of ongoing coevolution between the real and the fake. The peacock's tail, the lion's mane, the $20 bill, and the Rolex are not static designs. They grow ever more costly, precise, and elaborate over time as imitators try to reap the social, sexual, and status benefits of such displays without possessing the underlying qualities being displayed (fitness, health, wealth, or taste).

## Signaling, Branding, and Profit

If you want to make a decent profit, your product must have a special signaling value beyond its nominal function. If a product appeals to everyone, it cannot signal anything about the consumer, so consumers will simply comparison shop for it on the basis of features and/or price. Neoclassical economics assumes this is what consumers do, but it is the last thing that real businesses want from consumers, because it drives profits toward zero.

In actual capitalism, corporations strive mightily to avoid competition based on mere objective product performance. Instead, they use advertising to create signaling systems—psychological links between brands and the aspirational traits that consumers would like to display. Although these signaling links must be commonly understood by the consumer's socially relevant peer group, they need not involve the actual product at all. The typical *Vogue* magazine ad shows just two things: a brand name, and an attractive person. It is irrelevant whether the person is wearing any of the brand's clothing. Mere clothing can be copied within a few weeks by any coastal Chinese sweat shop. Often, the ad contains no text other than the brand name, no price information, no product features, no retailer locations—seemingly nothing that could guide a rational consumption decision.

However, there is a hidden rationality at work—the rationality of costly signaling. What matters in most advertising is the learned association between the consumer's aspirational trait and the company's trademarked brand name—the fountainhead of all profitability.

Often, celebrity endorsements are the easiest way to create such an association: the celebrity's traits can be linked to the product brand without the traits themselves needing to be identified explicitly. For example, Mont Blanc pen ads that feature Johnny Depp or Julianne Moore can create a mental link between the Mont Blanc pen and these actors' widely recognized and admired traits (coolness, attractiveness, intelligence, sense of humor, emotional authenticity)—without having to name those traits as such. These ads also mention Mont Blanc's support of the Entertainment Industry Foundation's National Arts Education Initiative, so a further association with generosity and creativity is established. A similar logic drives the Burberry ads featuring Kate Moss and its contributions to the Breast Cancer Research Foundation. In short, celebrities are portrayed in ads not just for their name recognition, but for the distinctive traits they are believed to have, and these become associated, through the symbolic magic of classical conditioning, with the product itself. Since celebrities are not widely known for their generosity, reinforcement of the status signal by the flying buttress of corporate philanthropy helps consumers feel better about conspicuous consumption.

The ad viewer himself need not believe that the brand has any logical or statistical link to the aspirational trait he wants to display. He must simply believe that other ad viewers from his social circle will perceive such a link. If I want to look tough, I don't need to believe that the Hummer H1 really looks tough; I need only believe that more gullible onlookers will think it looks tough, and will credit me with toughness for owning it. Thus, all ads effectively have two audiences: potential product buyers, and potential product viewers who will credit the product owners with various desirable traits. The more expensive and exclusive the product, the more the latter will outnumber the former. Thus, most BMW ads are not really aimed so much at potential BMW buyers as they are at potential BMW coveters, to induce respect for the tiny minority who can afford the cars. This explains why BMW sometimes advertises in mass-circulation magazines: it is an inefficient way to reach their target market of potential BMW buyers but it is a very efficient way to reach the BMW coveters who might respect the BMW buyers. Their true target market recognizes this fact, because they, too, sometimes read mass-circulation magazines, and see that their less-successful peers are being educated to understand the semiotic power of the BMW 550i. This is how any signal bootstraps its way from arbitrary association into common knowledge.

Advertisers can make errors when they do not understand this signaling logic. De Beers has recently begun advertising diamond rings for single professional women, by trying to introduce a new social convention: whereas traditional engagement rings are for the left hand, these single-woman rings are "right-hand rings." At first, this sounds great: men used to buy diamond rings for their fiancées when they got engaged, but today there are many wealthy working women who are not engaged, and who might nonetheless like a diamond ring. However, signaling theory suggests this campaign may be counterproductive. If unengaged women start buying themselves diamond rings, and observers don't bother distinguishing right from left ring-fingers, then diamond rings will no longer reliably display that a loving man has spent two months' salary on a woman. The diamond's signaling power will evaporate; it will no longer advertise a woman's attractiveness, agreeableness, happiness, and faithfulness, but only her earning power. Worse, the new synthetic gem Moissanite is also advertising to single women as an inexpensive, undetectable alternative to diamond, and Moissanite marketers aren't even encouraging women to wear their rings only on the right hand. If unengaged women are going around with $300 Moissanite rings indistinguishable from $30,000 diamond rings, engaged couples will question the signaling value of the diamond ring. They may switch to opal nose studs.

## Why Bother Signaling?

Costly signaling theory became important in biology not just because it solved some technical problems about how signals can stay reliable over evolutionary time. It also became significant because it clarified the diverse and profound benefits of signaling across many species. It showed not only how signaling could work, but, more important, why it's worth doing. If an animal can credibly signal its individual qualities to others, that can bring several key benefits.

First, one's quality signals can solicit parental care. Young animals that credibly signal their prospects for surviving and reproducing can solicit more parental care, feeding, and protection. This benefits the young animal by reducing its chances of dying young, and promoting its healthy, safe development, and it benefits the parents by allowing them to allocate their limited time, energy, and food to offspring that are most likely to pass on their genes. This pattern, called "discriminative parental solicitude," has a dark side, in that parents tend to neglect or even kill young animals that display conspicuous cues of genetic inbreeding, birth defects, stunted growth, poor health, or behavioral incompetence. Perhaps this is why human children try to display their physical and mental competencies by doing difficult things while screeching "Hey, Mom, look what I can do!" They evolved to act as if they knew that such displays may be rewarded by fitness-promoting forms of parental investment, such as cookies. Children whose physical or mental defects preclude such conspicuous quality signals—those with Down syndrome, autism, or congenital blindness, for example—are subject to much higher rates of parental abuse, neglect, and homicide. Since human maturation is especially slow, and human parents live especially long, parental investment often continues to be important throughout young and even middle adulthood. Thus, offspring continue to display their qualities unconsciously to their parents, by graduating from college, marrying well, becoming law-firm partners, having cute, healthy children of their own, and baking them cookies. Children who showed early promise, but who subsequently betrayed their reproductive prospects by becoming death-row inmates, often provoke some withdrawal of parental affection and investment, however unfair and discriminatory such decisions may be.

Quality signals can also be used to solicit care and investment from other genetic relatives. Because relatives share overlapping sets of genes, they are shaped by evolution to act as if they have overlapping interests. This is called kin selection, and just as parents have incentives to allocate their parental solicitude to the most "deserving" offspring (those most able to convert parental care into future reproductive success), relatives allocate their familial solicitude to the most deserving kin. (Indeed, to theoretical biologists, discriminative parental investment is just a special case of kin selection.) Thus, the healthiest, most attractive individuals in an extended-family clan tend to elicit the greatest attention and fondness from their relatives. They get more cookies from grandmothers and more job offers from uncles. From this viewpoint, family reunions can be seen as periodic rituals for mutual quality displays among genetic relatives: each individual tries to display his or her physical and mental traits in the best light to potential familial benefactors, and at the same time tries to assess which relatives are worthy of receiving his or her generosity. Poor families may have public-park barbecues while rich families congregate at estates in Kennebunkport or Balmoral, but in each case, similar social functions are served. Privileges, hopes, expectations, and resources are redistributed according to quality inspections of newborns, marital-prospect assessments of juveniles, and longevity assessments of the elderly. We all want to look worthy to our relatives, to the extent that they can do anything for us.

Moreover, quality signals can also be used to solicit social support, alliances, and friendships from nonrelatives. This is an important tactic for all animals who live in social groups larger than their kin group, who can recognize individuals, and who can support or ignore them discriminatively based on previous quality assessments and interactions. Among social primates like us, such relationships are critical for individual survival and reproduction. Popular apes live long and prosper; ostracized apes end up dead and childless. So, we have evolved irrepressible instincts to display our individual qualities to any potential supporters, allies, or friends who can offer us social benefits. This is the most ancient form of charisma-based politics, and the root of cliqueishness and clubbiness. Sometimes its benefits are abstract, delayed, and indirect: young-adult popularity yields midlife business contacts. But often, especially in prehistory, its benefits were dramatic, immediate, and direct: local celebrities are first protected and last abandoned under conditions of warfare, starvation, or illness. Even Achilles was better defended by his fierce Myrmidons than by his allegedly invulnerable skin.

Finally, quality signals can attract and retain sexual partners—the very gateways to reproductive success. This is a key factor for all species in which females or males have some power of mate choice. Mate choice can have profound effects in shaping quality signals, but for now, suffice it to say that if the opposite sex is choosy about its sexual partners, then one has extreme incentives to display one's qualities both to opposite-sex potential mates and to same-sex rivals. Those who display most impressively will attract the highest quality and quantity of mates, and deter the highest quality and quantity of rivals. This quality-signaling process is absolutely central to evolution in most sexually reproducing species, including our own. Even if one survives to a ripe old age through signaling one's excellence to parents, kin, and friends, one is an evolutionary dead-end if one does not attract at least one sexual partner.

These four modes of signaling often overlap: the same traits that show off one's physical and mental health to parents and kin can also attract friends and mates. Beauty and sanity are broadly valued.

Beyond these four modes of signaling, an individual animal can also benefit from quality displays by using them to deter potential predators from chasing it, to deter potential parasites from attacking it, and to deter rival groups from attacking one's own group. In each case, one doesn't have to convince the predator, parasite, or hostile group that one could never be overcome, only that some other victim would be an easier target. Quality signaling to predators and parasites is not a central concern for modern humans in developed nations, since we rarely encounter packs of wild hyenas, baboons, or mosquitoes. However, collective quality signaling to potentially hostile groups is the essence of gang warfare, interethnic rivalry, and international politics. Conspicuous consumption at this collective level plays a central role in quality signaling between human groups. Nations compete to show off their socioeconomic strength through wasteful public "investments" in Olympic facilities, aircraft carriers, manned space flight, or skyscrapers. While such prestige goods may sometimes work to attract foreign investment and tourism, and to deter military encroachment, it can also turn irrational. For instance, the United States' determination to signal its military and economic superiority through a $3 trillion war in Iraq seems to have induced a massive recession that threatens its long-term status as a superpower.

The four main reasons for displaying individual qualities—to solicit parental care, kin investment, social friends, and sexual partners—favor many of the same traits. Your parents, kin, friends, and mates all attend to your physical and mental health, because this influences the likelihood that you will survive into the future to yield the fitness benefits of their attentiveness. They all care about your physical attractiveness, because they unconsciously realize this influences the likelihood that you will attract good mates to pass along their genes (if parents or kin), or make them look good by association (if friends), or pass along their attractiveness to your joint offspring (if mates). They all care about your intelligence, because this influences your prospects of both survival and reproduction; your social value as a relative, friend, or mate; and your genetic value as a mother or father. They all care about your personality traits and moral virtues, because these influence your likelihood of being kind, fair, and conscientious in any social role you play. So, most traits that are valued in one type of social relationship are also valued in other types, and this is why we strive to display those traits consistently to different social audiences.

## Signals of Body and Mind

My analysis focuses on products that signal the key human traits— bodily traits of health, fitness, fertility, youth, and attractiveness, and mental traits of intelligence and personality. In some ways, these two different classes of traits represent different levels of description for the same human phenotype, the same individual organism. The body is the physical phenotype as it appears to others perceptually—through vision, hearing, touch, taste, and smell. We use products such as clothing, makeup, cosmetic surgery, and exercise equipment to modify the body's appearance so it seems healthier, younger, fitter, more fertile, and more desirable. At a higher level of abstraction, the mind is just what the body does with itself—the behavioral phenotype as it appears to others, who will make judgments about our intelligence and personality. Just as we aspire to make our physical phenotypes (bodies) look better, we do the same for our behavioral phenotypes (minds, personalities), by spending money on education, charity, travel, hobbies, and bumper stickers to appear smarter, kinder, more outgoing, and more open-minded than we may really be. At a still higher level of description, an individual's status, prestige, position, popularity, fame, and wealth constitute his social phenotype as it is perceived by others, and as it emerges through a lifetime of socializing, conversing, trading, friend making, coalition building, and status seeking. We spend money on luxuries and status symbols to appear more reputable, popular, and rich, but because these social traits emerge fairly directly from our physical and mental traits, *Spent* will consider them in the context of the first two levels.

For both physical and mental traits, we have an interest in flaunting our fitness—in overstating our true, stable, personal qualities—so others will treat us better as friends, lovers, relatives, or colleagues. We rarely admit to this, yet we almost always notice when others are doing it. If you're a Columbia graduate like me, you may think of yourself as a serious, hardworking, socially conscious, urbane intellectual. But if your sexual rival went to Harvard, you may dismiss him as a pretentious, social-climbing, hypocritical, narcissistic fake. And vice versa. When we flaunt our own traits, we're just playing the self-presentation game effectively. But when we realize we've been duped by fake signals from others—would-be lovers, friends, politicians—we view them as cheats and liars. Deep, wide, and thick is our self-deception about signaling.

This self-deception makes it hard to be a fully conscious consumer. We're seldom honest with ourselves about why we buy things, and advertising euphemisms don't help. Which slogan sounds better: "L'Oréal: Because you're worth it," or "L'Oréal: Because you want to look younger than the skanky Starbucks barrista who's always flirting with your husband"? How about these: "The 2006 BMW 550i: Poised for performance," or "The 2006 BMW 550i: Poised to leave burning tire smoke in the spotty faces of those Subaru WRX-driving punks who threaten your masculinity as a divorced 47-year-old orthodontist." The true emotions and aspirations behind such purchases must not be revealed, lest we realize that we're trying to buy things that can't be bought—or that aren't worth the cost. Consider: the 2006 BMW 550i goes from 0 to 60 mph in 5.4 seconds and retails for $57,400; the Subaru WRX STI goes 0 to 60 mph in 4.7 seconds and retails for $32,445. Is the BMW badge really worth $25,000? You could buy a replacement BMW badge for $16 from Autopartswarehouse.com, glue it to your Subaru, outrace the other orthodontists, and still cover your divorce lawyer's fees. Or, you could cover three years of weekly psychotherapy for self-esteem and anger-management issues.

By now it should be clear that you'll be most comfortable with my arguments if you fully accept yourself as a fitness-flaunting consumer narcissist who has been deluded, throughout your whole life, into irrational spending habits by advertising euphemisms and peer pressure. In other words, you'll probably feel uneasy for much of the time you're reading it. The truth is, science sometimes hurts.

## Conspicuous Consumption as Fitness Signaling

People have radically diverse responses to the very idea of conspicuous consumption. Some folks consider it blindingly obvious that most human economic behavior is driven by status seeking, social signaling, and sexual solicitation. These include most Marxists, marketers, working-class fundamentalists, and divorced women. Other folks consider this an outrageously cynical view, and argue that most consumption is for individual pleasure ("utility") and family prosperity ("security"). Those folks include most capitalists, economists, upper-class fundamentalists, and soon-to-be-divorced men. Such differences of opinion can rarely be resolved by trading examples or anecdotes, or arguing from first principles. It more often helps to apply some psychology. So, inspired by costly signaling theory, my colleagues Vladas Griskevicius, Josh Tybur, and others ran a series of four experiments whose goal was to see how people's consumption decisions might shift as the potential mating benefits of costly signaling became more or less salient.

In the first experiment, college students came to the lab in small groups. Each was randomly assigned to one of two conditions: "mating" or "nonmating." The mating subjects looked at three photographs of attractive opposite-sex people on a computer screen, picked which one they thought was most desirable, and spent a few minutes writing about an ideal first date with that person. The nonmating subjects looked at a street scene photograph and spent the same amount of time writing about the ideal weather for walking around and looking at the buildings it featured. Then, all subjects were asked to imagine that they had a modest windfall of money (such as a lottery win of a few thousand dollars), and to choose which of several conspicuous luxuries they would want to buy (such as a new watch, European vacation, or new car), as opposed to saving the money in a bank account.

They were then asked to imagine that they had some extra time available per week, and were asked to choose how many hours they would spend volunteering (such as working at a homeless shelter or helping at a children's hospital). The results were dramatic: men in the mating condition said they would spend much more money than men in the nonmating condition (for example, they might take the European vacation rather than saving that money), but there was no mating effect on women's consumption decisions. On the other hand, women in the mating condition said they would spend much more time volunteering than women in the nonmating condition, but there was no mating effect on men's volunteering. This study confirmed that conspicuous consumption (for men) and conspicuous charity (for women) can be increased by thinking about mating opportunities, and so can function strategically as a form of mating display.

Because costly signaling theory suggests that signals must be conspicuous and publicly observable in order to attract friends or mates, my colleagues wanted to see whether this mating effect applied especially to conspicuous rather than inconspicuous consumption and volunteering. In a second experiment, another set of college students were randomly assigned to similar mating or nonmating conditions. Then, subjects indicated how much money they would want to spend on the same conspicuous consumption luxuries (new watch, European vacation) from study 1, or on some new "inconspicuous" necessities (such as basic toiletries, kitchen staples, household cleaning products). Finally, subjects indicated how much time they would want to spend on the same conspicuous volunteering from study 1, or on some inconspicuous but socially helpful activities (such as picking up trash alone in a park or taking shorter showers to conserve water). The results here were equally clear: men in the mating condition, compared with the nonmating condition, said they would spend more money on the conspicuous luxuries, and that they would actually spend less on the inconspicuous necessities (household cleaning products); there was no effect on female consumption decisions. By contrast, women in the mating condition, compared with those in the nonmating condition, said they would spend more time on conspicuous pro-social volunteering (such as working at the children's hospital), but no more time on inconspicuous pro-social activities (such as taking shorter showers); there was no effect on male volunteering. So, thinking about mating does not simply increase overall consumer spending or pro-social volunteering; it only increases conspicuous consumption or conspicuous charity—the behaviors that work best as public, costly displays.

It was a bit surprising that in both studies, the mating-primed men did not act more conspicuously benevolent, and the mating-primed women did not spend more on conspicuous consumption. Maybe mating-primed men only favor conspicuously heroic forms of benevolence (such as saving strangers from drowning), and mating-primed women only favor conspicuously generous forms of spending (such as bidding high at charity auctions). So, in study 3, another set of students followed the same routine as in study 2, except that they could choose to spend money on the original forms of conspicuous consumption (such as the new watch or car), or on more generous forms of conspicuous spending (such as donating to natural disaster victims at an on-campus booth, bidding high at a public auction to raise money for sick children). Also, they could choose to spend time and energy on the original forms of conspicuous charity (such as working in a homeless shelter), or on some more heroic activities (such as running into a burning building to save someone trapped, distracting a grizzly bear from attacking a stranger). As predicted, mating-primed women compared with control-condition women said they would spend more on generosity-signaling conspicuous spending; mating-primed men did the same. Also, mating-primed men compared with control-condition men said they would do more heroic helping, but not more nonheroic helping; there was no effect of mating condition on female heroic helping. Moreover, men who were most interested in promiscuous, short-term sexual liaisons showed the largest increase after the mating prime in both generosity-signaling conspicuous spending and in heroic benevolence. This is especially strong evidence that men are using these behaviors as costly mating signals.

If thinking about mating can increase men's heroic benevolence, perhaps other kinds of male benevolence might be boosted by mating motives—not just heroic acts, but charitable activities that also allow men to display their dominance or leadership. In study 4, a final set of students, mating-primed or not, indicated how willing they would be to do helpful things that were either low-status (the original five activities from study 1), or socially prestigious (volunteering with Hollywood celebrities in the Make a Wish Foundation for terminally ill children, or coordinating meetings between charities and White House officials), or socially dominant (giving a speech for a good cause to a hostile crowd, or leading a risky public protest). Both sexes showed a marginally higher interest in socially prestigious pro-social behaviors when they were mating-primed. However, only the men showed a higher interest in the socially dominant pro-social behaviors when they were mating-primed, and this effect was carried mostly by the highly promiscuous men who are most motivated by mating effort.

A fascinating recent paper by Jill Sundie, Vladas Griskevicius, and their colleagues replicated these effects in four further studies. Inspired by this finding that highly promiscuous men are most influenced by mating primes, they measured interest in short-term mating using a scale called the "sociosexuality inventory." Study 1 showed that high-promiscuity men were more willing to borrow fashionable clothing from a friend to impress a potential mate rather than a new boss, whereas low-promiscuity men would rather impress the boss, and women showed no difference. Study 2 showed that high-promiscuity men who looked at photos of eight attractive women, compared with those who looked at photos of eight attractive buildings, said they would spend more money on items that were rated by other students as examples of conspicuous consumption (such as designer sunglasses or an elaborate car stereo) rather than inconspicuous products (such as low-cost jeans or a toaster oven). There was no mating-prime shift for low-promiscuity men or for women, and the standard questionnaire for measuring "materialism" did not predict conspicuous consumption. Study 3 showed that the mating-prime effect on conspicuous consumption only works when the potential mating situation is a short-term hookup rather than a long-term relationship—and even then, it only works for the high-promiscuity men. Study 4 showed that women rated a man driving a Porsche Boxster as more attractive for a short-term sexual relationship than a man driving a Honda Civic, but the Porsche did not make the man more attractive as a possible marriage partner. Men rating women were uninfluenced by the type of car she drove. This last study is especially intriguing, since it suggests that women are attracted to conspicuously consuming men for their good genes (which can be obtained from a single copulation) rather than their good resources (their wealth as it would be relevant in a long-term marriage).

A final study by the evolutionary psychologists Margo Wilson and Martin Daly confirmed that mating primes influence economic behavior more strongly among males than females. They were interested in people's "discount rates," which determine how patient people are given a choice between a certain number of dollars tomorrow, or a larger number of dollars a larger number of days into the future. First they measured the discount rate for about two hundred subjects, using standard economic-choice measures. Then they asked people to look at photographs of potential mates or cars that were previously rated as highly attractive or unattractive. Finally, they remeasured each person's discount rate to see if it had changed after looking at the photographs. They found that men who looked at the highly attractive photographs of women (from Hotornot.com) switched to a much higher discount rate—they became much less patient about money. Looking at cars had no effect on men's discount rates, and looking at men had no effect on women's discount rates. (However, women looking at highly attractive cars actually developed a lower discount rate— a more economically rational attitude better suited to saving up the money for buying such a car.) In short, men who saw attractive women became much more motivated to get whatever money they could in the short term, presumably so they could spend it on conspicuous consumption to attract mates.

These nine studies nicely support my key point: much of human economic behavior, whether consumption or charity, is engendered by motives of costly signaling to display our personal qualities to potential mates and other social partners. These motives are finely tuned and very specific. They show systematic sex differences, and are influenced by apparent mating opportunities. Among mating-primed people, they especially provoke conspicuous rather than inconspicuous behaviors. Among mating-primed women, they especially provoke charitable spending rather than luxury spending. Among mating-primed men, especially promiscuous men, they provoke heroic, socially prestigious, and socially dominant forms of pro-social behavior. Such finely patterned behaviors seem unlikely to arise as a side effect of general excitement or arousal. They reveal a human display psychology with intricate design features shaped over millennia of evolution, to attract mates and friends through certain kinds of costly, risky behaviors that reliably signal certain desirable traits.

# 7

# Conspicuous Waste, Precision, and Reputation

COSTLY SIGNALING THEORY yields a usefully paranoid perspective on human behavior, for it leads us inevitably into doubt. It induces a vivid skepticism about other organisms' claims of bodily, mental, and social superiority. It makes life more unnerving—less like a formulaic romantic comedy, and more like a combination of film noir, spy thriller, and science fiction. Ultimately, it provokes more than a nodding agreement with the observation by nineteenth-century American minister Elias Root Beadle, that "half the work that is done in the world is to make things appear what they are not."

While we have instincts for ostentatious self-display, we also have instincts for optimality and frugality, qualities that guided our ancestors' choices of how to hunt and gather efficiently, trade profitably, and make tools skillfully. Whereas Thorstein Veblen famously analyzed our self-display instincts in *The Theory of the Leisure Class* (1899), he also examined our efficiency-seeking instincts in his lesser-known *The Instinct of Workmanship and the State of the Industrial Arts* (1914). Those efficiency-seeking instincts can also be applied to evaluate different signaling systems. They lead us to think that we would like to be able to judge other people accurately, without their having to handicap themselves so onerously, by wasting so much matter, energy, time, and risk on signals. Is there any escape from this nightmare of costly signaling, wherein the only alternative to universal deception is colossal waste?

For a few years in the early 1990s, biologists thought there might be an answer to this dilemma in the theory of "indexes." Indexes were, theoretically, signals that could be perfectly reliable without being costly: the quality of one organism's underlying trait would correlate

100 percent with the apparent quality of a cue being observed by other organisms, so the cue would be a reliable index of the trait. Indexes might be reliable by tapping into some fundamental constraint of Newtonian physics or developmental biology. For example, an animal's adult body size seemed like a possible index of its genetic quality, bodily condition, and age, because mutated, sickly, newborn animals simply can't eat enough quickly enough to grow larger-than-average bodies. In this view, a four-centimeter-long guppy must be a pretty good guppy, if guppies average three centimeters; a twenty-five-meter-long brachiosaurus must have been a pretty good brachiosaurus, if brachiosaurs averaged twenty meters.

However, index theory was eventually overtaken by life-history theory—the study of how organisms allocate their resources, energy, and time to different patterns of growth and behavior across their life spans. Life-history theory suggested that there was always plenty of room for faking, even when it came to allegedly reliable indexes. Suppose animals evolve to pay more attention to one another's body length as a cue of underlying quality when trying to choose the fittest mates or trying to invest resources in the fittest offspring. There will then be selection of mates and offspring to allocate more energy to growing longer body parts, to get more than their fair share of copulations or parental care. The body parts that are cheapest to grow longer, because they have the smallest cross sections (necks and tails), will lengthen most quickly over evolutionary time. Thus, there will be a gradual reduction in the correlations between apparent body size (length) and actual genetic quality, bodily condition, and age. The result will be long-tailed guppies and long-necked, long-tailed brachiosaurs. Empirical tests of life history theory suggest that such evolutionary reallocations of energy from one growth pattern to another growth pattern are rather easy to achieve. There are not, in fact, many hard "developmental constraints" on organisms that can guarantee the reliability of any particular index. Evolution is just too potent and clever at circumventing constraints, and turning reliable indexes into unreliable advertisements. This same inexorable undermining of index reliability happens at the cultural level, as illustrated by the counterfeiting of paper money, luxury watches, and Rembrandt paintings.

It would be convenient to live in a world of reliable indexes and merit badges. We could judge books by their covers, people by their faces, and products by the simplest of visible cues. Our children could prosper by learning a few key proverbs: the bigger, the better; no pain, no gain; like father, like son; might is right; practice makes perfect. They wouldn't need to learn that all that glitters is not gold, or that actions speak louder than words. They could keep their innocence and gullibility forever, in a world of eternal, ubiquitous truth.

Costly signaling theory is basically the bad news that we do not live in that world, because the survival and reproductive incentives for deception are too high, and the adaptive processes that invent new forms of deception—genetic evolution, cultural invention, individual improvisation—are too fast and pervasive. The good news is that some signal reliability is possible, under certain conditions of signal cost.

Signal cost can certainly include costs in terms of an animal's general energy budget (food calories eaten) or ecological resources (such as mating territories acquired by threats and fights), which are similar in some ways to money. When such moneylike costs are paramount, costly signaling theory most resembles Thorstein Veblen's theory of conspicuous consumption. In *The Theory of the Leisure Class*, Veblen argued that luxury goods and services are acquired by the rich mainly to display their wealth, not to increase their happiness. Buyers of top-of-the-range products understand that their high price is a benefit, not a cost. It prevents poorer buyers from owning the same product, thereby guaranteeing the product's reliability as an indicator of their possessor's wealth and taste. The rich covet the new iPod not for the sounds it can make in their heads, but for the impressions it can make in the heads of others. Veblen applied the term "conspicuous consumption" to all such costly displays, wherever the main function of acquiring, using, and displaying the product is to signal one's individual traits, wealth, and status to observers. Costly signaling theory simply generalizes Veblen's insight to the biological world. It observes that animals, including humans, often show off the most expensive signals they can afford, whether those signals are peacock tails or Hummer H1s. In each case, reliable signaling demands some sort of "conspicuous waste"—a highly visible expenditure of resources that brings no material benefit, but that simply signals the expender's ability and willingness to waste those resources.

However, the term "cost" can be misleading, and is not limited to biological analogues of monetary cost. Signal costs can also include costs in terms of an animal's time (minutes spent doing something), attention (proportion of consciousness allocated to some ongoing task), diligence (proportion of quality-control systems invested in growing or producing a well-formed signal), physical risk (probability of injury or death), or social risk (probability of embarrassment or punishment if fakery is discovered). In these cases, costly signals may not demand conspicuous waste; they may demand instead conspicuous precision (which can be achieved only through time, attention, and diligence) or conspicuous reputation (vulnerability to social sanctions).

Signals that guarantee their reliability through conspicuous waste or conspicuous precision can be called indicators. If they reveal an animal's genetic quality or phenotypic condition, they are fitness indicators— they reveal that animal's fitness, its statistical propensity to survive and reproduce successfully under a species' natural ecological conditions. The peacock's tail is a fitness indicator that relies partly on conspicuous waste (it is large and heavy), and partly on conspicuous precision (its finely grown feathers are radially symmetrical at the macro level, show evenly spaced eyespots at the meso level, and display iridescence at the micro level).

Most human-designed products show some combination of conspicuous waste and conspicuous precision. The Hummer H1 SUV is heavy on conspicuous waste (it weighs 7,850 pounds and gets about ten miles per gallon), but shows very little conspicuous precision—indeed, it shows only enough precision of design and manufacture to run for a while before it needs a repair (which happens fairly often, according to the *Consumer Reports* reliability ratings). The Lexus LS 460 sedan is somewhat lighter on conspicuous waste (it weighs 4,240 pounds and gets about twenty miles per gallon), but much heavier on conspicuous precision (fit, finish, features, reliability, luxury).

Apart from indicators, many animals evolve "badges" of fitness or status. These do not rely on conspicuous waste or precision, but on conspicuous reputation. Badges may be easy to grow and maintain for those who are socially recognized as deserving them. But if individuals display them without meriting them, those individuals are subject to social punishment by others, ranging from avoidance and ostracism to physical harassment and mob violence. For example, facial markings act as status badges among female paper wasps. Higher-status wasps have a larger number of dark spots above the mouth, and spot number correlates reliably with head width, body size, and fighting ability. The biologists Elizabeth Tibbetts (University of Michigan) and James Dale (Simon Fraser University) found that if they painted extra dark spots on lower-status wasps, those wasps were attacked much more aggressively by dominant wasps. Thus, the dark spots may be metabolically cheap to produce, but lead to heavy social punishment if the wasp does not have the fighting ability to back up its status badge. Analogous human badges include British school ties, military medals, and gang tattoos—in each case, unmerited display provokes sanctions, ranging from ostracism to death.

For badges to work as reliable signals, they must be checked periodically by other animals to determine if the badge wearer really merits the badge. This could be construed as a "maintenance cost" of the badge, but it is a cost borne not so much by the badge wearer as by the audience. Others must take the trouble to check for fraud, forgery, and cheating, and must impose sufficient punishment on cheaters that low-status individuals can expect to pay a prohibitive cost for trying to wear a high-status badge. This leads to another problem of reliability: detecting and punishing cheaters is inconvenient to the individual (since it costs time, energy, and risk), but good for the group (since it keeps the badge reliable), so it is technically an act of altruism. Such pro-social altruism can be maintained in populations if the badge checkers and cheater punishers gain social and reproductive benefits from policing the signaling system. But the whole system relies on a complex balance of power and interest between honest badge wearers, cheats, police, police admirers, and police groupies. If a population gets the balance right, badges can be extremely reliable, efficient, and cheap as signals of fitness, or status, or anything else. But if individuals forget to reward badge checkers and cheater punishers with extra social status, friendship, and mating opportunities, the system can easily break down.

Most human-designed products also rely to some degree on status badges, which are called brands. Brand names and logos are usually light on waste (they're small and flat), and only moderate on precision (they use fairly simple fonts and designs), but they are ferociously protected as badges by trademark law, corporate lawyers, and cargo inspectors. Even in China most fake branders have learned to alter the imitated brand name just enough to avoid outright trademark violations, so they produce Paradi (rather than Prada) clothes, PenesemiG (rather than Panasonic) batteries, and Pmua (rather than Puma) shoes. As with the randomly patterned black dots on wasp heads, the precise form and meaning of the brand name or logo are immaterial. Even stupid brand names (Accenture, Babolat, Bong Vodka, Intellocity, Kork-Ease, and Phat Farm) can succeed if actively promoted and policed. All that matters is that the brand is recognizable to consumers, is positively correlated with some aspect of product quality, and is enforced by legal and social sanctions. Just as evolution builds badges as signaling systems by spreading the genes for growing and recognizing the badges, and enforcing their accuracy, corporations build brand equity by promoting consumer recognition of the brand, positive association with the brand, and intolerance for faking of the brand. The same recognition, association, and reliability considerations apply to corporation and conglomerate names above the brand level (Toyota Motor Corporation), and to product names within each brand (Camry).

The table below illustrates the differences between conspicuous waste, conspicuous precision, and conspicuous reputation as signaling principles.

| basis of comparison | conspicuous waste | conspicuous precision | conspicuous reputation |
|---|---|---|---|
| form of cost | matter, energy | attention, skill | cheater punishment |
| form of signal quantity | mass quality | information brand | recognition |
| typical cues | large size | small tolerances | large sales |
| | costly materials | accurate design | distinctive design |
| | surface area | symmetry | prototypicality |
| | transience | reliability | fashionability |
| | scale | intricacy | popularity |
| displayer emotions | largesse | pride | vanity, conceit |
| audience emotions | awe | fascination | familiarity, envy |
| terms of praise | fancy, fun | fine, fit | famous, fashionable |
| elements | gold (watch) | silicon (chip) | neon (sign) |
| foods | pâté de foie gras | sushi | prime rib |
| watch brands | Franck Muller | Skagen | Rolex |
| clothes | sable coat | Issey Miyake dress | Armani suit |
| car brands | Hummer | Lexus | BMW |
| house features | great room, foyer | kitchen, garden | facade, postal code |
| education degrees | Oxford M.A. | MIT physics Ph.D. | Harvard M.B.A. |
| cities | Los Angeles | Singapore | Paris |

Each example is arguable, because most real products rely on all three principles in varying proportions. In particular, most successful products must display some minimal level of precision in order to function efficiently, safely, and credibly for their nominal purpose. The Hummer must run at least as long as the test-drive lasts; Los Angeles must have streets sufficiently well-organized that commuting to work takes no more than twenty-four hours per day. However, none of this precision needs to be conspicuous precision—it does not need to go beyond the level of pragmatic necessity. Also, most products must bear some recognizable branding, or else they would not command any price premium above a generic commodity, and they would not deliver any profit to their manufacturers, distributors, and retailers. Conspicuous waste is also optional if other signaling principles are in play. Some products assert their quality through an almost pure mix of conspicuous precision and branding, without extravagance of scale or materials. For example, most Skagen brand watches use very thin, simple cases of stainless steel, precise quartz movements, and discreet branding.

Of course, conspicuous waste, precision, and reputation do not exhaust the possible forms of signal reliability. There is also conspicuous rarity: pink diamonds, Rembrandt paintings, moon dust, Princess Diana dresses. Rarity is so valued by exotic pet enthusiasts that whenever a new species is described in the scientific literature, the species' extinction risk is increased by collectors seeking out the high-premium exotica—as happened with the turtle *Chelodina mccordi* from the Indonesian island of Roti, which brought up to $2,000 on the international pet market and then almost went extinct. There is also conspicuous antiquity (often correlated with rarity): Roman Double Aureus gold coins, Muromachi period samurai swords, Gutenberg Bibles. However, these more exotic forms of signal reliability apply mostly in specialist auction markets for unique luxury items, not in mainstream consumer-product design and marketing.

## The Relative Efficiency and Morality of Different Signaling Systems

Each signaling principle has its distinctive pros and cons from the viewpoint of the signaler, the audience, and the population and ecology at large. These distinctions are significant but often overlooked.

For example, socialist and environmentalist critiques of runaway consumerism apply most forcibly to cruder forms of conspicuous waste, which sequester matter and energy for the rich at the expense of the poor, and which impose the largest ecological footprint (resource and energy requirements). It is much harder to raise socioecological objections to an iPod nano than to an H1 Hummer. Aristocrats differ from the nouveaux riches not in their freedom from consumerism, but in their preference for conspicuous precision and reputation ("the finer things in life") over conspicuous waste ("the crass and vulgar"). Green-minded, dreadlocked vegans differ from well-coiffed soccer moms not in their aloofness from capitalism, but in the forms of conspicuous reputation they prefer: Nature's Path Organic Ginger Zing Granola Cereal with Silk Plus Omega 3 DHA soy milk, rather than Kellogg's Frosted Flakes with factory-farmed milk full of growth hormones and mastitis-induced cow nipple pus. A Scientologist who spends $280,000 on the cult's "intensives" and courses to achieve "Operating Thetan Level VIII" status may consider himself spiritually superior to a Beverly Hills psychiatrist who spends that amount on a Girard-Perregaux Magistral Tourbillon Swiss watch, but both are chasing conspicuous reputation, just in somewhat different forms. Thus, arguments about consumerist capitalism can go far astray when we do not recognize that there are many different forms of reliable signaling—and our own favored signaling tactics are the ones we are least likely to recognize as signaling at all.

In many ways, conspicuous waste is the simplest, most popular, and (surprise!) most wasteful form of signaling. Guppies and brachiosaurs managed to evolve wasteful signals, and the senses and brains to discriminate between them; so, too, have humans. Indeed, conspicuous waste is a metabolic extension of life itself. Biologists define life as a set of processes that can sustain reproduction (self-replication) and metabolism (control over local matter-and-energy flows to preserve one's bodily adaptations in the face of entropy). Conspicuous waste is simply a way to display the scope of one's control over those local matter-and-energy flows, by monopolizing more resources than are necessary for short-term self-preservation. The payoff of such signaling is that it promotes success in the longer-term self-preservation of oneself (through higher social status) and one's genes (through higher mating success). For example, an active adult male human needs about three thousand calories per day. An actively status-seeking billionaire entertaining thirty guests and twenty staff on his three-hundred-foot megayacht might control the flow of an additional 150,000 calories per day for feeding his dependents, plus 8,000 gallons of fuel per day if cruising at fifteen knots. A gallon of fuel contains about 30,000 calories of energy, so the megayacht's total energy budget is about 240 million calories per day—equal to 200,000 pounds of porterhouse steak, or 80,000 times the calories needed by the billionaire's own body. A skilled prehistoric hunter would have been very lucky to bring home 40 pounds of meat in a day, so the billionaire is demonstrating control more than five thousand times the metabolic resources that any normal human could command throughout human evolution. Thus, the yacht's food and fuel budget is a prodigious extension of the billionaire's metabolism.

Thomas Malthus observed that human populations usually expand to match the environment's carrying capacity. Given a Malthusian world of limited resources, one man's monopolization of such massive energy flows through a yacht means that other people do without. So, when poor, hungry people see megayachts and other conspicuous waste, they tend to get upset, and they either demand socialist revolution (in the nineteenth century), or better antidepressants (in the twenty-first). Veblen's *Theory of the Leisure Class* was a satirical tirade against conspicuous waste, but it did not explore the psychology of why we find such waste so aesthetically offensive. His 1914 book, *The Instinct of Workmanship*, filled that gap by positing that humans evolved, over millennia of tool-making and technical innovation, a deep instinctive preference for efficient tools, projects, and lifestyles. In Veblen's view, aesthetic revulsion against conspicuous waste reflects this instinct of workmanship, which values precision of design and efficiency of function. Veblen envisioned a technocratic utopia in which righteous engineers drove corrupt marketers and investors from power, and delivered right-size, minimal-waste products to a grateful public.

However, in signaling-theory terms, this instinct of workmanship could be construed as simply a different set of signal preferences—the preferences that we apply when judging conspicuous precision rather than conspicuous waste. To a large degree, Veblen's technocratic utopia was achieved throughout much of the twentieth century across the developed world, through the aesthetics of international modernism, minimalism, and techno-fetishism. These abandoned the conspicuous waste of Victorian ornamentation for the conspicuous precision of design, form, and functionality—exemplified by Frank Lloyd Wright houses, Knoll furniture, Movado watches, and Apple computers. The rich still bought costly furniture, but it was now costly because designers spent hours trying to develop new forms of chairs that required machinists to invent new fabrication methods for each novel design, rather than because wood-carvers spent hours chiseling Rococo Revival floral motifs into mahogany.

The twentieth century's shift from conspicuous waste to conspicuous precision was beneficial in many ways. It empowered designers to explore more creatively the space of possible product designs. It minimized, for a while, the reputability of grotesque Gilded Era extravagance. It paved the way for the eco-aesthetics of "small is beautiful." It shifted status from the engineers of the very large (trains, battleships, skyscrapers) to the engineers of the very small (electronics, biotech, nanotech). It increased consumer appreciation of fit, finish, reliability, functionality, and novelty. While conspicuous waste continues to be favored in a few places (the United States, Russia, Saudi Arabia), conspicuous precision has become more fetishized elsewhere (Japan, Korea, Hong Kong, Europe).

However, conspicuous precision can lead to equally absurd forms of runaway consumerism. If the basic functions of consumer goods remain stable, and if form follows function, then modernist design should have quickly settled on optimal designs for every product category—whether it is the ideal chair, car, or house. Moreover, if consumers demand conspicuous precision, as manifest in perfect workmanship, quality, and reliability, then every product should operate efficiently for many decades, if not lifetimes, passed down the generations as capital legacies that parallel the biological legacies of our genes. Such a system would, however, have resulted in economic catastrophe, because eventually no one would have to produce or buy anything. This was the businessman's nightmare of the 1950s, and it was solved—with a great deal of explicit strategizing among investors, marketers, and politicians—by inventing the various forms of planned obsolescence and technological pseudoprogress that journalist Vance Packard examined memorably in books such as *The Hidden Persuaders* (1957), *The Status Seekers* (1959), and *The Waste Makers* (1960).

Corporations realized that if they wanted to continue selling new cars, and that if modernist aesthetics dictated that each of a car's functions could be served by only a narrow range of optimal design forms, then the only way to "improve" each year's new model car was to incorporate novel functions that arise through perpetual technical innovation. We are all familiar with the parade of improvements that rendered each car model "obsolete" within a few years, including: air-conditioning (1941), power windows (1948), power steering (1951), cruise control (1958), three-point seat belts (1959), and so on. More recently, anxious drivers who were once happy with the driver-side airbag (introduced in 1980) now feel obligated to upgrade successively with passenger-side airbags (1987), side airbags (1995), knee airbags (1996), windowbags (1998), and second-generation airbags (1998). My 1997 Toyota Land Cruiser has only two airbags, so I feel irrationally vulnerable compared with the driver of a 2007 BMW 750Li, which has eight airbags—although my Land Cruiser weighs more and rides higher.

Thus, consumerist signaling through conspicuous precision drives very fast proliferation in product features and functions, often through minor technical innovations and pseudo-innovations. The resulting waste is not usually obvious within any single product, but is serious when one adds up all the successively obsolete products that a consumer buys within a lifetime. If you buy a Toyota Prius (2,900 pounds curb weight) every three years rather than a Land Cruiser (5,700 pounds) every ten years, then over a fifty-year driver's lifetime, you'll consume a total of 48,000 pounds rather than 28,000 pounds of car mass. Continual upgrades don't always make sense, even when the new product has lower environmental costs per mile driven. Similarly, a consumer of my generation will have consumed (so far) a couple of record players, an 8-track player, several cassette players, many CD players, and various iPods—each made with enough conspicuous precision that they would probably still work, if I could find them.

The shift from conspicuous waste to conspicuous precision reflects a gradual dematerialization of consumption, by which we signal superiority through design, not mass, and through intricacy, not size. Mobile phones are already becoming too Lilliputian for adult males to use without feeling like a palsy-pawed giant ground sloth. The twenty-third generation iPod nano of 2045 might be a cubic millimeter of literal nanotech glued to one's earlobe. If it falls off, we'll have to be careful not to breathe it in, or to let our pet Shih Tzu lick it off the floor. Conspicuous precision is quickly reaching the limits of our visual acuity and fine motor control.

Yet this dematerialization of products is somewhat illusory, in that we have simply centralized the allocation of capital from the products themselves to their engineering and manufacturing facilities. A 100-million-transistor PC microchip may be only one square centimeter, but it can be made only in a billion-dollar factory with a million square feet of floor space, including 100,000 square feet of "Class 1" ultra-clean area. To achieve conspicuous precision in our products, we pay companies to cover the capital investments in manufacturing capacity and expertise that are implicit in a product's quality, rather than through paying them to invest extravagant capital in each product itself. The result can be marvelous economies of scale, since the centralized manufacturing capabilities can churn out many products at very low marginal costs once they are up and running.

Conspicuous reputation represents an even more extreme demate-rialization of consumption. In this realm, a product's signaling reliabil-ity no longer depends on the capital invested in the product itself (as in conspicuous waste), or in the product's design and manufacturing (as in conspicuous precision), but in the product's marketing and brand-ing. The product's reputability and the brand's equity exist not in the product's material form, but in the brains of consumers and observers. Those brains are just as real as steel or silicon, but because they can-not be manipulated as directly as ordinary matter can be, they must be reached through the senses: through advertising, product placement, opinion leaders, imitation, word of mouth, and all the other arma-ments of modern marketing.

In 2006 Interbrand reported that Coca-Cola had $67 billion in global "brand equity"—more than any other company. To nonmarketers, the concept of brand equity (the total value of a brand's name-recognition among consumers) usually sounds so abstract as to be meaningless. To marketers, however, it has very real empirical meaning—namely, that 94 percent of people on earth recognize "Coca-Cola," and most "respond positively" to its products, which means that they will pay a price premium for a Coca-Cola product over a physically equivalent generic beverage. Given a global population of 6.5 billion people, Coca-Cola's mind share is worth about $10 per person, on average; that is, the company's $2 billion of advertising per year over many decades has gen-erated product recognition propensities that are literally worth about $10 per human brain. The other top ten brands—Microsoft, IBM, GE, Intel, Nokia, Toyota, Disney, McDonald's, Mercedes—each have 2006 brand equity greater than $21 billion, or more than $3 of product rec-ognition propensity per human brain. These propensities may seem elusive, immaterial, even mystical, but they are real enough to support the careers of hundreds of thousands of marketing and advertising pro-fessionals. No doubt as the new science of neuroeconomics progresses and uses ever more sophisticated brain imaging methods to identify which parts of consumer brains respond to brands and products, con-cepts such as brand equity will seem ever less mystical.

Costly signaling theory highlights the fact that brand equity exists mostly in the minds of signal receivers (observers of other people's product consumption), not in the minds of signalers themselves (actual consumers of a product). The luxury brands with the highest brand equity (Louis Vuitton, Gucci, Chanel, Rolex, Hermès, Tiffany, Cartier, Bulgari, Prada, Armani) understand this perfectly well. They advertise in *Vogue* and *GQ* not so much to inform rich potential consumers that they exist, but to reassure rich potential consumers that poorer *Vogue* and *GQ* readers will recognize and respect these brands when they see them displayed by others. This is why the typical luxury ad includes a highly attractive model dressed up as a high-status heiress, wearing an expression of contempt and disdain for the viewer. The ad does not say "Buy this!"; it says, "Be assured that if you buy and display this product, others are being well trained to feel ugly and inferior in your presence, just as you feel ugly and inferior compared with this goddess."

Conspicuous reputation as a signaling principle provokes its own distinctive moral and aesthetic objections. Critics of branding point not so much to each product's material wastefulness or its techno-fetishistic proliferation of useless features and functions, but to its invidious social-comparison effects. Branded products lead the consumer to feel higher in status, sexiness, or sophistication—feelings that are ultimately either oppressive (if observers grant higher status to the product displayer, and thereby feel inferior) or self-deluded (if observers do not actually grant the higher status). In either case, the branding seems iniquitous—a waste of human effort, attention, and vanity in the zero-sum game of social status.

However, in other respects, conspicuous reputation is wonderfully efficient as a signaling principle. It leaves a very modest ecological footprint, because it relies on lightweight information and media technologies to influence people's minds. Advertising is just photons aimed at eyes, and sound waves aimed at ears. These can be produced through very efficient media—print, radio, television, Internet—that do not burn much fossil fuel or club harp seals to death. The principle of conspicuous reputation honestly acknowledges the core function of consumerist capitalism—the invidious display of one's personal qualities to observers—and so allows marketers and consumers to fulfill that function with less collateral damage to other people, and environments.

It seems unlikely that people will ever relinquish their runaway quest for self-display, as the failures of communism and hippie utopianism showed all too clearly. (Note that Mikhail Gorbachev of the USSR and Keith Richards of the Rolling Stones are now both appearing in ads for Louis Vuitton luggage.) Yet, people's modes of self-display are quite flexible, as shown by the development of different display norms across different historical epochs and cultures. So, self-display may one day be shifted from our current antisocial, irresponsible, unreliable forms of conspicuous waste, precision, and reputation to more pro-social, conscientious, reliable forms that still let people make a living.

# 8

# Self-Branding Bodies, Self-Marketing Minds

THE BODY IS a practical tool for survival and reproduction, but it is also the packaging and advertising for our genes. Body watchers have understood for centuries that the human form, like that of all animals, is a showcase of fitness indicators. By understanding how we display good genes, good health, and good fertility through our body traits, we can better understand how our self-marketing minds display themselves through our consumer behavior.

Folk wisdom holds that beauty advertises health and fertility. Publilius Syrus, the ancient Roman mime, wrote, "A fair exterior is a silent recommendation." Oscar Wilde, the gay Irish wit, wrote, "It is only shallow people who do not judge by appearances." Martha Graham, the angular modernist dancer, observed, "The body never lies." Peacocks have tails, lions have manes, and humans have luxuriant head hair that will grow long if left uncut, gemlike eyes with big white scleras, expressive faces, sensually everted lips, smooth hairless skin, and gracefully dextrous hands. As Darwin realized, these human traits are all the result of sexual selection—generations of our ancestors favoring reliable cues of youth, health, fitness, and fertility in their sexual partners. Moreover, each human sex has its own luxury fitness indicators. Males have beards, large jaws, large upper-body muscles, and longer, thicker penises than those of other great apes. Females have much-enlarged breasts and buttocks and relatively thinner waists compared with those of other great apes. These sex-specific indicators only mature at puberty, just in time for individuals to advertise their fitness in the mating market.

In the past fifteen years, evolutionary psychologists have confirmed that many of these human body traits do function as fitness indicators. (That might sound obvious, but it's not always easy to find good empirical support for the allegedly obvious.) There is now strong evidence that these body traits are uniquely amplified in humans, are valued as sexual signals, and are displayed more prominently during sexual courtship. They show large sex differences, and develop under the influence of sex hormones, especially around puberty. Some of these traits (lips, breasts, buttocks, penises) even get engorged with blood during sexual arousal. They are salient and sexy, they are costly and complex, and they are hard-to-fake signals of survival and reproductive ability. They are the focus of visual pornography for males, and narrative pornography for females. As we age, sicken, or starve, our hair dulls, faces droop, lips thin, skin sags, and hands shake. If male, our penises, muscles, and jaws shrink; if female, our breasts and buttocks shrink and sag. If our parents were siblings or cousins, their genetic inbreeding reduces our health, fertility, and attractiveness. If we are badly injured, the damage reduces our symmetry of form and grace of movement. These traits are both physical fitness-indicators, and physical-fitness indicators (the hyphens matter here).

These indicators are most reliable at the medium to low end of fitness. Superattractive movie stars like Anne Hathaway and Will Smith are not necessarily healthier or more fertile than the most attractive 20 percent of your high school class. Yet both groups are very likely healthier and more fertile than the least attractive 20 percent of your high school class. Attraction to physical beauty is the flip side of repulsion to physical ugliness, asymmetry, disease, handicaps, lesions, and injuries. When we buy beauty-enhancing products, we are mostly trying to hide imperfections—the sorts of conditions discussed on Embarassingproblems.co.uk: acne, bad breath, belly-button discharge, cold sores, dandruff, genital infections, head lice, impotence, incontinence, jock itch, penis problems, snoring, tics, varicose veins, and warts.

Indeed, when we think of the phenomenal range of human bodies, it's very hard to remember what we have in common, for only the fitness indicators are salient. If you compare the actress Elizabeth

Hurley with your ugliest female relative, or the soccer star David Beckham with your ugliest male relative, it's hard to imagine that their kidneys, colons, tongues, eyeballs, and testicles or ovaries are virtually identical in structure, function, and physiology. If you're one of the 8 million people who saw the "Body Worlds" exhibition of plastinated human corpses, by Dr. Gunther von Hagens, you may remember that you couldn't really tell which body had been sexually attractive before the skin flaying, muscle dissection, and organ disassembly. If you were one of the 3,400 people executed by China in 2004, your sex appeal had little influence on the price that your kidneys, corneas, liver, and heart could fetch on the medical black market. Some rich Singapore businessman is probably looking happily now through those condemned corneas, without prejudice.

In the cadaver trade (for medical schools and surgical training seminars), the big companies such as Innovations in Medical Education and Training (IMET) don't care how cute you were before your corpse entered the "tissue bank" supply chain. They'll just pay the standard $550 for a head, $815 for a whole leg, $1,500 for a spine, or $375 per breast—plus air freight. When the occasional package bursts open in a FedEx sorting plant, the workers don't exclaim "Oh, what a fetchingly handsome jawline and noble brow on that disarticulated human head!"; to them, it's just a repulsive ball of thawing meat, fat, and bone. As Augie Perna, the head of IMET, said in a *Harper's* interview, "That torso that you're living in right now is just flesh and bones. To me, it's product." Or, as the Stoic philosopher Epictetus wrote nineteen hundred years ago, "You are a little soul carrying around a corpse."

I don't want to rock your existential boat with such morbid thoughts; I just want you to remember that overall, human bodies are really pretty similar, and that we make an enormous fuss over what are essentially minor differences. To a short, malnourished, rickets-warped, lice-infested medieval peasant, we would all look as gorgeous as angels. To a horny chimpanzee, we would all look like anemic, short-armed, sunken-faced chemotherapy patients. To a saber-toothed cat, we would all look mighty tasty. Members of the International Biometric Industry Association have to work quite hard to find reliable automated ways to distinguish people. Typically, this requires measuring precise, complex, semirandom surface patterns from the skin (fingerprints, hand scans, spectral reflectance profiles), face, iris, or voice. To other species and biometric machines, human differences are subtle indeed.

But the evolutionary consequences of those differences become huge when they are averaged across whole populations and compiled across thousands of generations. Tiny differences in DNA sequences lead to subtle differences in proteins, cells, tissues, organs, and bodies, which lead in turn to different average levels of reproductive success. We evolved to notice these apparently trivial differences because in the evolutionary long run, they matter enormously.

Obviously, many goods and services are marketed to improve physical appearance. The most popular and dramatic examples—clothing, cosmetic surgery, the fitness industry—have been analyzed by thousands of researchers. Only the more extreme oddities still surprise, and remind us how much narcissism lurks beneath consumerism. One such example is Bust-Up chewing gum, marketed to Japanese schoolgirls by B2Up, a Tokyo-based company. It contains phytoestrogens claimed to increase breast size, shape, and firmness (see Bustgum .com). Another example is *Extreme Makeover,* a reality TV show in which volunteers are filmed undergoing liposuction, breast implants, nose jobs, tummy tucks, exercise regimens, and wardrobe upgrades.

## The Rise of the Triathlon

When the jogging craze swept the developed world in the 1970s, runners started to compete in marathons. While the Boston marathon had been run since 1897, big new annual marathons began sprouting up everywhere: New York (1970), Berlin (1974), Chicago (1977), London (1981). The New York marathon, the world's largest, now has to limit the runners to thirty-seven thousand per year.

However, from a fitness signaling viewpoint, marathon-running developed a problem: it proved too easy. It soon became apparent that almost anybody between fifteen and fifty-five could finish a marathon, if he or she trained conscientiously for about six months. Completing a 26.2-mile marathon was no longer an achievement special enough to provoke much respect, and winning a marathon became impossible for the nonprofessional runner. Even worse, marathon training did not give people very impressive bodies. Long-distance runners become skinny and stringy: men lose their upper-body muscles, while women lose their breasts, buttocks, and fertility. Both sexes develop lower sex-hormone levels and libidos. Finally, marathon-running was too inexpensive to serve as a good wealth indicator: all you needed to compete was a pair of $80 running shoes.

Some other form of amateur athletic competition was needed—one that would favor stronger, fatter, sexier, more fertile bodies, and that required much costlier equipment. Voilà—the Hawaiian Ironman Triathlon was invented in 1977. It required a 2.4-mile swim, followed by a 112-mile bike ride, and ended with a 26.2-mile marathon. This contest favored amateur athletes with more upper-body strength (for swimming), lower-body strength (for cycling), and fat reserves (to avoid freezing in the water, and to provide long-term energy). Today, serious triathlon competitors also need to spend at least $5,000 on equipment—for example, $550 for a Blue Seventy Helix Full triathlon wetsuit, $3,800 for a Scott CR1 Plasma Pro triathlon bike, plus special energy bars and electrolyte drinks (a triathlon, unlike a marathon, requires refueling en route). So, triathlons quickly became more reputation enhancing than marathons: longer, harder, and more competitive, requiring more muscle, more training, and more money. Insofar as ancestral human life also required a balance between the aerobic capacity for long-distance walking and running, and the strength to carry hunted meat, gathered plant-foods, and heavy babies, the triathlete's body also better fits our evolved sexual preferences for the way that a healthy, fertile human should look.

The triathlon is a classic example of runaway trait display. It not only required a higher level of physical fitness, but also higher levels of wealth, training effort, and conscientiousness. Compared with the marathon, it also resulted in a body better suited to human sexual preferences for general health, strength, and fertility. The triathlon's displacement of the marathon exemplifies a key signaling principle: strong signals drive out the weak.

## Facial Fertility Indicators and Cosmetics

One of my central claims is that we buy many products in a semiconscious attempt to increase the social and sexual attractiveness of certain personal traits. There is little mystery about some of the physical traits (breasts, buttocks, hair) that function as fitness indicators, and that women display to appear sexually attractive. However, many other physical traits function as more-subtle fitness indicators in each sex, and are more often displayed in public, and augmented by purchased products. Women's cosmetics, for example, teach us a lot about the nature of beauty: how certain female facial traits function as indicators of fitness, youth, and fertility.

The evolutionary background of cosmetics is that in most primate species, sexual selection focuses very heavily on facial appearance. This is because most primates are highly social (they care where others are looking and what their facial expressions are conveying), and highly visual (they care what others' eyes and faces look like). Typically, this leads female primates to select male primate faces that have some absurdly exaggerated sexually attractive features. The whole face itself may evolve to be more visually salient, as in the strong light-and-dark contrasts that frame the faces of black and white colobus monkeys, white-faced sakis, and De Brazza's monkeys. The eyes may evolve to be more salient, with similarly stark contrasts framing the eyes, as in the white-nosed monkey, red-capped mangabey, and spectacled langur. The facial skin or fur may evolve highly saturated colors, as in the uakari's bright red face and the mustached monkey's bright blue face. These colors may evolve to be slightly more subtle, as in the pastel blue of the golden snub-nosed monkey, or the pastel pink of the Yunnan snub-nosed monkey—or the colors may evolve to be even more garish, as in the mandrill's bright red and blue nose. Other primates evolve striking hairstyles, such as the emperor tamarin's mustache, and the cotton-top tamarin's mohawk. Facial features may be exaggerated to form oversize noses (the proboscis monkey), cheek flaps (the orangutan), or pointy crania (the gorilla).

In each case, male primates have the more distinctive and extreme appearance, reflecting an evolutionary history of stronger female choice on their facial fitness indicators. In humans, by contrast, sexual selection has affected each sex's facial features about equally, resulting in distinctively male "testosteronized" features (more prominent brows, jaws, chins, and noses; deeper-set, smaller eyes; beards) that are strikingly different from female "estrogenized" features (larger, more prominent eyes; fuller lips; lighter, smoother skin).

Many cues of female facial attractiveness function not just as general fitness indicators that reliably reveal a woman's stable genetic quality, but as fertility indicators that reveal whether a woman is at the peak of fertility in her life course and in her ovulatory cycle. Female humans are unusual in that they become sexually active in their mid-teens, several years before they are physiologically fertile (which, given low-fat ancestral diets, was usually the late-teen years); and they remain sexually active long after they have stopped being physiologically fertile (after menopause). They are also unusual in remaining sexually active throughout their menstrual cycles, rather than mating just when they are at peak fertility, before ovulation. By having sex when they are not really fertile, female humans can solicit more consistent help, care, and investment from males for themselves and their children. This, however, creates a problem for males: If they have limited resources to invest in themselves and in different possible girlfriends and alleged offspring, how do they choose the investment opportunities with the highest likelihood of promoting their reproductive success? That is, how do they discriminate the truly fertile women who could get pregnant from the juvenile, postmenopausal, menstruating, or already-pregnant women?

In assessing women's ages, men apparently evolved to pay close attention to facial and bodily cues of being in the young-adult phase of peak fertility. Women's fertility increases rapidly in the late teenage years, peaks in the mid-twenties, and declines smoothly throughout the thirties, reaching negligible levels in most women by the early forties. Thus, prehistoric men who felt sexually attracted to pre-fertile girls or post-fertile women might have had wonderful, loving relationships, but they left few descendants to inherit their maladaptive preferences. Prehistoric men who felt most sexually attracted to peak-fertility women in their late teens through mid-thirties left more descendants to inherit their preferences for cues of youthful but sexually mature fertility.

As an evolutionary counterstrategy, women evolved to develop their fertility cues at younger ages in their juvenile life stages, and to retain their fertility cues into older postmenopausal ages, in order to extract more attention and investment from male mates. In prehistory, there were no birth certificates, driver's licenses, or birthday parties to yield objective indexes of a woman's chronological age. Males had no way to assess actual age or fertility; they had only visual and behavioral cues of sexual maturity, youthfulness, and fertility. So, women could evolve to fake their fertility all the way from around age twelve to around age sixty—not just physically but behaviorally. As a counter-counterstrategy, men in turn evolved more perceptive ways to distinguish genuine peak fertility from pseudofertility—and to react with sexual impulsiveness to the genuine-fertility cues. For instance, when Bill Clinton allegedly had an affair with Monica Lewinsky in 1995, Monica (born 1973) was twenty-two, near peak fertility, whereas Bill's wife, Hillary Clinton (born 1947), was forty-eight, with negligible fertility.

One way of faking fertility across a broader age range is to apply cosmetics that amplify facial fertility cues that peak in young adulthood, such as plump lips, large eyes, prominent cheekbones, smooth and radiant complexion, thick and glossy head hair, and minimal facial hair. (Each of these traits distinguishes not only peak-fertility females from younger or older females, but also human females from other anthropoid apes.) In almost all cultures, women use cosmetics to make their facial traits look more similar to those of a mid-twenties peak-fertility woman.

For example, eye size, whiteness, and contrast tend to decline after the mid-twenties fertility peak, so women use eyeliner, eye color, mascara, brow pencils, and Visine to make their eyes look larger and clearer, and to increase the light/dark contrast between pupil, iris, sclera, lashes, skin, and brows. (Virtually all magazine covers, ads, and porn pictorials also use Photoshop image manipulation to whiten the scleras of female models.) The thickness, redness, and eversion of lips tends to peak in the mid-twenties as a fertility indicator, so women use lip liner, lip color, lip gloss, and lip plumper to make their lips look larger and brighter. The prominence and roundness of "cheekbones," which are really the pads of estrogenized fat above the zygomatic bones, tend to shrink with age, and the overlying skin loses its vasodilated red blush, so women use blush and shading to highlight cheekbones. (Many models and actresses also get small silicone pads implanted in their upper cheeks, and fat extracted from their lower cheeks, to make their cheekbones more prominent.) The translucence, radiance, evenness, and smoothness of women's facial skin declines after the mid-twenties, so women use foundation and powder to make the skin complexion appear more uniform in pigment and texture, and to hide wrinkles. The thickness, length, color saturation, and glossiness of head hair peaks in the mid-twenties, so older women use volumizing shampoos, shine-enhancing conditioners, hair dyes, highlights, hair extensions, and wigs to emulate peak-fertility hair. Conversely, they use depilators to minimize facial hair, which tends to grow as estrogen drops and androgens increase after menopause.

Top makeup artists, such as Pat McGrath of CoverGirl, have the skills to achieve these youth- and fertility-enhancing effects with such subtlety that casual observers cannot tell any cosmetics have been used. The result is that when Molly Sims or Queen Latifah appear in CoverGirl ads, they look so youthful, radiant, and fertile in such undetectably natural-looking ways that most female readers feel instantly outclassed, and desperate to achieve the same hyper-fertile look. (The beauty-fertility connection is not usually conscious, in the minds of either women seeking to be more beautiful or men seeking women who are more beautiful. We have just evolved to act as if we understand the connection.)

Cosmetic choices are much less culturally arbitrary than they appear at first glance. Ancient Egyptians may have used kohl rather than liquid eyeliner to increase apparent eye size, red ochre rather than blush to increase cheek redness, and mesdemet (ground copper and lead ore) rather than foundation to make their complexions look more uniform. Yet in each case, they sought to increase rather than decrease the facial cues of estrogenization, youthful sexual maturity and fertility. Across cultures, people have used different cosmetic ingredients, pigments, colors, and bases, and different cosmetic application methods, styles, and patterns. Yet I cannot find any cases in which cosmetics have been widely used by women in a culture to give an impression of small, jaundiced eyes, pale thin lips, or wrinkled, pockmarked skin.

The fact that all cosmetics aim for the same youth- and fertility-enhancing effects makes it very difficult to be a cosmetics marketer or product developer, because it is so hard to capitalize on genuine functional innovation in the cosmetics themselves. Pharmaceuticals that genuinely "cure diseases" can be profitably patented after FDA approval, but cosmetics that merely "enhance appearance" are not subject to the same evidentiary standards or legal protections. Instead, cosmetics brands must differentiate themselves mainly through packaging and pricing, to appeal to women who differ along the dimensions of age, wealth, and apparent sexual availability (if not actual promiscuity). For example, the Wet'n'Wild brand appeals to low-status, self-sexualizing teens, whereas the Shiseido brand appeals to high-status young professional women seeking a chaste image that deters sexual harassment at work. Recently, cosmetics have also begun brand differentiating according to more subtle lifestyle dimensions: conventional romanticism (Chanel, L'Oréal, Lancôme) versus clinical chastity (Clinique, Olay, Neutrogena, Prescriptives), versus eco-friendly naturalism (the Body Shop, Aveda, Ecco Bella), versus metrosexual urbanism (DuWop, NYX, Smashbox, Urban Decay). Women will pay a high price premium to acquire the brand that they feel best expresses their personality—although the chemical differences between the products are negligible, and although most potential male mates have no idea that the squat, squarish, black, recyclable tubes of Ecco Bella lipstick are supposed to connote environmental awareness, whereas the cylindrical, brassy tubes of Urban Decay lipstick are supposed to connote cutting-edge, post-goth glamour.

## From Signals of Bodily Fitness to Signals of Mental Fitness

The cosmetics analysis above could be misread as suggesting that women display more fake physical fitness cues than men do. In fact, a similar analysis could be done of male fashion models, Chippendale's dancers, athletes, bodybuilders, soldiers, policemen, ruggedly hirsute field anthropologists, and users of anabolic steroids, Viagra, and Rogaine. Vanity about physical appearance is an equal-opportunity vice; the males just target different physical traits for amplification and display, using different products. For instance, men try to build muscle mass by buying ultrapremium sustained-release protein powders such as Syntha-6, advertised with close-ups of muscular male torsos that are hotter than most of those found in gay porn. Both sexes draw on the costly signaling principles of conspicuous waste (enlarged cheekbones, hair volume, breasts, and buttocks in women; increased height, muscle volume, and sexual performance in men) and conspicuous precision (facial and bodily symmetry, regularity, and smoothness; minimal wrinkles, lumps, and lesions on the skin; fine grooming and clothing).

And yet many consumers wind up disappointed with products that promise to enhance their physical appearance. They realize that youth, health, fertility, and fitness are actually very hard to fake, because people have evolved for thousands of years to be very discerning. Our perceptual systems evolved the greatest sensitivity and accuracy in the tasks that were most important to our social and reproductive success, and assessing others' physical qualities were among the most critical of them. We may not be able to see at a glance which cantaloupe in the produce section is ripe (hence more nutritious and carrying fewer phytotoxins), but we can see which potential mate in the nightclub is "fit" (hence more fertile and carrying fewer genetic mutations).

Intelligent adults eventually realize all this, at some level. They stop fooling themselves that body-display products actually increase physical attractiveness, and learn instead that maintaining one's physical appearance is an effective way of broadcasting one's personality traits. The consistent, skillful use of cosmetics, razors, hair products, and fashion advertises one's intelligence, mental health, conscientiousness, and self-esteem. The fortysomething trophy wives of sixtysomething movie producers know that they cannot really look better on a purely visual, sexual level than the twentysomething aspiring actresses. But they do know subconsciously that, by maintaining a svelte figure, subtly made-up face, and strong fashion sense, they can remind a husband that they still have sufficient savvy and self-respect to make a useful ally in parenting and networking, or a formidable opponent in divorce court. Likewise, the ex-college-quarterback who maintains his form and energy after twenty years of marriage is displaying not so much his physicality as his moral self-restraint against the temptations of sloth and gluttony. Thus, even the apparently superficial use of appearance-enhancing products can, among mature consumers, signal a wide array of mental and moral traits, rather than just bodily traits of health, fertility, or fitness.

## Looking Tough in World of Warcraft

Consumer narcissism can be seen most clearly in the virtual worlds of massively multiplayer online games (MMOGs), which are great natural experiments for investigating human behavior. For example, about 9 million people globally play a MMOG called World of Warcraft (WoW), marketed by Blizzard Entertainment. In WoW, players create a virtual character who is one of ten races (such as human, dwarf, gnome, orc, or blood elf), and one of nine classes (such as warrior, shaman, rogue, or hunter). The characters run around in a three-dimensional virtual world forming teams, fighting monsters, trading items, and accumulating character levels (which range from 1 to 70) depending on amount of game experience. Online players can see one another's characters rendered in very high detail and vivid color, so the visual appearance of one's character and the items it owns (especially its weapons and armor) are a major source of pride and social status within the gaming community.

My colleague Zack Mendenhall analyzed conspicuous consumption in WoW by looking at the auction-market prices of different weapons. Each weapon has a set of game-relevant attributes determined by the game designers, such as the "damage per second" it typically inflicts, its "durability" (the time until it breaks), and its "critical strike bonus" (its chance of inflicting an extra-damaging attack). While these quantitative attributes wholly determine each weapon's actual utility in the game, some weapons look a lot cooler than others, and are more impressive to other players. These appearance-based features of the weapons are indexed by their "rarity" color level, which can be white (common), green (uncommon), blue (rare), or purple ("epic," or very rare). For example, purple swords look larger, more exotic, and more distinctive than green swords. They are also objectively better weapons (with greater damage per second, durability, and other useful qualities), but we wondered whether weapon rarity in itself might be a source of social status for players, and whether they might pay a price premium for rare weapons.

In online chat associated with the game, players often discuss by name the rare, high-status weapons that they own, want, or have observed. For example, in online WoW dialect, they might say:

"Whodi, 411 flash. u c that l70 pally in SM with Cataclysm's Edge?"
"NFW. IIRC, it does 138 dps? WEG. Better than my vendor trash"
"=:–D. OMG, sick wep, sick tank! If he got torqued, I would shit a cold purple Twinkie. *g* L8R"

Standard-English translation:
"Friend, consider this interesting news. Did you see that level-70 paladin running around the Scarlet Monastery with that famous purple-level sword, Cataclysm's Edge?"

"No feasible way. If I remember correctly, that sword does 138 points of damage per second? (Wicked evil grin expression.) Better than my cheap, common weaponry."

"(Scared grin expression.) Oh my God, what a cool weapon, and a cool, physically intimidating warrior! If he got angry at me, I would react with extreme distress. (Giggles.) Talk to you later."

Mendenhall and I focused on the simplest "melee" weapons (swords, axes, and daggers) that cause damage at short range, and on the green, blue, and purple rarity levels that are favored by almost all players. We analyzed just those weapons that could be auctioned to other players through the WoW online market for an auction price determined by supply and demand, and that could also be sold back to a nonplayer character in the game for a vendor price determined by the game designers. This yielded a sample of 309 specific weapon models, each of which is owned by thousands of different players around the world. (We excluded two superweapons with special hard-to-quantify powers and astronomical auction prices—the Night Blade dagger and the Blinkstrike sword.) We expected that conspicuous consumption effects would lead to weapon rarity's having a much bigger effect on a weapon's auction price (which reflects the premium players will pay for cool-looking weapons) than on its vendor price (which reflects a weapon's objective attributes).

This is exactly what we found: rarity has a far larger effect on a WoW weapon's auction price than any of its game-relevant attributes, explaining 33 percent of price variation. Median auction prices were 157 silver pieces for green weapons, 2,923 for blue, and 11,099 for purple. By contrast, a weapon's vendor price is 95 percent explained by damage per second, and not at all by rarity (median vendor prices were 90 for green, 120 for blue, and 467 for purple—differences that became insignificant when controlling for damage per second). Thus, WoW players are willing to pay a premium of about ten thousand silver pieces for purple-rarity weapons that look big, cool, and exotic, even controlling for the better quantitative attributes of those weapons. Such coolness displays work because most rare weapons are acquired not through auction, but through being a high-status expert player who has spent hundreds of hours going on successful quests with large, competent guilds of friends.

The advantage of analyzing conspicuous consumption in such virtual worlds is that there is no wiggle room for arguing that the luxury good has some hidden quality or performance benefit that justifies its price premium. BMW may claim that its luxury-car premium reflects hard-to-quantify aspects of handling and drivability. No such claims can explain the purple-weapon premium in World of Warcraft, because the quantitative weapon attributes that are made publicly available by Blizzard Entertainment wholly explain the weapon's objective utility in the game. In this online game, we have caught people red-handed paying huge costs just to impress one another. My argument is that such effects are ubiquitous in the real world, too, and that the alleged hidden quality and performance benefits of luxury goods are typically illusory—just vague ways for consumers to rationalize their consumer narcissism.

## The Body Goes Mental

Most animals have very little behavioral control over their physical appearance. They can groom themselves to keep feathers or fur clean, but they cannot select a different species, sex, age, shape, color, or body texture. Ever since humans invented body ornamentation at least a hundred thousand years ago, however, we have been able to transform our bodies in ever more dramatic ways. Tribal peoples wear animal masks; British civil servants cross-dress; children play dress-up; the Florida elderly don toddler-bright colors. As people do more of their socializing through virtual-reality worlds such as World of Warcraft and Second Life, their visual appearance is becoming less constrained by their true physical characteristics, and more constrained by their psychological traits, such as aesthetic preferences and idealized self-images. Virtual-reality users will soon be able to create avatars that resemble a mini-Mao, a Botox syringe, a mantis-legged cantaloupe, a pearl necklace, Nigella Lawson, or the evil Archimandrite Luseferous from the Iain M. Banks novel *The Algebraist*. Such customized avatars will reveal nothing about the physical appearance of the users, but a lot about their psychology. They will demonstrate more forcefully than ever before that consumerism is not about owning material objects, but about displaying the sort of personal qualities analyzed in the next chapter.

# 9

# The Central Six

Humans normally share the same twenty-three pairs of chromosomes in each of about 50 trillion cells. We all have about 206 bones and 640 muscles. We all have about three pounds of brain and six pounds of skin. We all get to enjoy about 600 million breaths, then we die. We are all one species, yet individual differences loom large. We notice not just physical differences between one another, as discussed in the previous chapter, but psychological differences. This is because, as William James noted, "There is very little difference between one man and another, but what there is, is *very important*."

A century of psychology has identified six major dimensions of variation that predict human behavior and that are salient to us. These are the key individual differences that distinguish human minds. These are the mental traits that can be measured with good reliability and validity, that are genetically heritable and stable across the life span, that predict behavior across diverse settings and domains (school, work, leisure, consumption, and family life), and that seem to be universal across cultures and even across many animal species. If you know how somebody scores on each of these "Central Six" traits, you can infer a lot about his character, capabilities, virtues, and vices.

Also, as we'll see in the coming chapters, these six traits are absolutely central to understanding consumerism, because they are the fundamental traits that we try to display to one another through the goods and services that we buy.

**G** is for general intelligence, the first of the Central Six traits. It is also known as smarts, brains, general cognitive ability, or IQ. Shortly after Charles Spearman's key work in 1904, intelligence became the best-studied, best-established trait in psychology. Higher intelligence predicts higher average success in every domain of life: school, work, money, mating, parenting, physical health, and mental health. It predicts avoiding many misfortunes, such as car accidents, jail, drug addiction, sexually transmitted diseases, divorce, and jury duty. It is one of the most sexually attractive traits in every culture studied, for both sexes. It is socially desired in friends, students, mentors, coworkers, bosses, employees, housemates, and especially platoon mates. It remains ideologically controversial because its predictive power is so high, and its distribution across individuals is so unequal. Chapter 11 gives more detail about how intelligence is measured and what it predicts.

Over the past few years I have taken note of many bumper stickers on Albuquerque vehicles. They seem to advertise most extremes of the Central Six traits, and are useful in getting a sense of what these traits really mean. Here are some examples that seem to advertise higher- or lower-than-average general intelligence:

## High General Intelligence

- Talk nerdy to me
- A PBS mind in a Fox News world
- My border collie is smarter than your honors student
- Don't say ironic when you mean coincidental
- Sane, paululum linguae latinae dico (Sure, I speak a little Latin)
- Eschew obfuscation
- If it fits on a bumper sticker, it's not a philosophy

## Low General Intelligence

- Mommy says I'm special
- Collige
- TV is gooder than books
- Tongue pierthing ith thtupid

- I'm hung like Einstein and smart as a horse
- I wondered why the Frisbee was getting bigger, then it hit me

The other members of the Central Six are the Big Five personality traits: openness, conscientiousness, agreeableness, stability, and extraversion. These are more recent discoveries, dating back from around 1980, and they have revitalized the study of human personality. They clearly and efficiently map the main individual differences in human behavioral dispositions, and they are much more reliable and valid than other ways of describing human personality in psychology or marketing. Like general intelligence, the Big Five traits predict behavior rather well across different domains of life and different kinds of relationships. Like intelligence, they are genetically heritable, stable across life, and universal across cultures. Like intelligence, they are very salient when we choose social or sexual partners. They can be represented by the letters *O, C, A, S,* and *E,* and they can be reliably measured by several different personality questionnaires.

**O** is for openness to experience: curiosity, novelty seeking, broadmindedness, interest in culture, ideas, and aesthetics. Openness predicts emotional sensitivity, social tolerance, and political liberalism. People high on openness tend to seek complexity and novelty, readily accept changes and innovations, and prefer grand new visions to mundane, predictable ruts. They want to shine on like crazy diamonds. People low on openness tend seek simplicity and predictability, resist change, and respect tradition. They are usually more conservative, close-minded, conventional, and authoritarian. They don't mind feeling comfortably numb. High-O people tend to join strange new start-up cults, whereas low-O people tend to follow the same-old, better-established cults that their grandparents followed, which constitute the various organized religions. Openness correlates positively (but modestly) with intelligence, but also with certain types of mental illness, such as bipolar disorder ("manic depression") and mild schizophrenia ("schizotypy").

Here, again, are some real bumper stickers that seem to advertise different levels of openness:

## High Openness

- Question reality
- Legalize freedom
- My karma ran over your dogma
- Sorry I missed church. I was busy practicing witchcraft and becoming a lesbian
- Reality is where the pizza delivery guy comes from
- I like it sloppy and weird
- 1f u c4n r34d th1s u r34lly n33d t0 g37 l41d

## Low Openness

- Live it up, sinner
- Shut up, hippie
- Welcome to America. We speak English. Learn it or leave
- Stereotypes make life easier
- If God didn't want us to eat animals, he wouldn't have made them out of meat
- Gun control means using both hands
- What part of "Thou shalt not" don't you understand?

**C** is for conscientiousness: self-control, willpower, reliability, consistency, dependability, trustworthiness, and the ability to delay gratification. Conscientious people pursue long-term goals. They fulfill their promises and commitments, resist impulses and bad habits, and feel embedded in a network of mutual social obligations. In Chinese terms, they tend to build strong *guanxi*—a strong, reliable social network. They like to make plans, keep everything organized, seek perfection, crave achievement, and prefer doing one focused task at a time. People low in conscientiousness tend to be more comfortable with spontaneity and chaos. They accept things, people, and achievement levels that are "good enough" rather than "optimal," and they shift more easily among ongoing tasks. Yet they also show lower levels of drive and ambition. Conscientiousness predicts regular attendance at school and work, completion of assignments on time, cooperativeness in professional relationships, and civic engagement. It predicts eating a healthy diet, exercising regularly, avoiding drug addiction, and staying healthy. Along with intelligence, it is one of the two traits most desired by employers. On the other hand, because conscientiousness predicts effective contraception use, it is strongly disfavored by natural selection in the modern world. Extremely high conscientiousness shades over into obsessive-compulsive disorder and results in overwashed hands, whereas extremely low conscientiousness shades over into reckless impulsivity and results in long criminal records. Low conscientiousness is moderately related to one of the two main dimensions of mental illness: the "externalizing" dimension associated with childhood "conduct disorder" (delinquency), adult "antisocial disorder" (psychopathy), and "substance abuse" (alcoholism and drug addiction).

Some bumper stickers that advertise different conscientiousness levels:

### High Conscientiousness

- Get off the phone and crash already
- Jesus would have used his turn signals
- Just because you can, doesn't mean you should
- The harder I work the luckier I get
- A goal is a dream with a deadline
- The police never think it's as funny as you do
- Today is the tomorrow you forgot to plan for yesterday

### Low Conscientiousness

- Live every second as if your ass is on fire
- A clean house is the sign of a wasted life
- I'd rather be a failure at something I love, than a success at something I hate

- Saturday has a morning?
- Drive it like you stole it
- Oral sex is always a great last-minute gift idea

**A** is for agreeableness: warmth, kindness, sympathy, empathy, trust, compliance, modesty, benevolence, peacefulness. Saints are very agreeable; psychopaths are very disagreeable. People high in agreeableness tend to seek harmony, adapt to others' needs, and keep their opinions to themselves when doing so avoids conflict. People low in agreeableness tend to seek glory or notoriety, pursue their own needs, and express their opinions forcefully. Agreeableness is perceived not just as a personality trait, but as a moral trait. It corresponds to what most people would call "good" as opposed to "evil," "altruistic" as opposed to "selfish," "peaceful" as opposed to "aggressive." (Personality psychologists hate to sound judgmental about traits, so they try to avoid such moral terms.) Agreeable people often make much more pleasant long-term sexual partners, friends, relatives, in-laws, co-workers, and babysitters, so we often value agreeableness in others. In game-theory terms, the agreeable make better reciprocators, and contribute more to public goods, because they value other people's well-being ("subjective utility"), not just their own. Their motto is: "When in doubt, give." Disagreeable people, on the other hand, can often take social and sexual advantage of others, so there can be major evolutionary benefits to disagreeableness, especially for males. (This is why most wild animals are rather disagreeable, and why most humans, like other domesticated species, are much more agreeable.) People low in agreeableness can be cold, distant, aggressive, irritable, selfish, and arrogant; they lie, cheat, steal, rape, and kill more often. Their motto is: "When in doubt, take." Low agreeableness, even more than low conscientiousness, is related to the externalizing dimension of mental illness (delinquency, psychopathy, alcoholism, drug addiction), and with various nasty behaviors that impose high costs on others (promiscuity, philandering, wife beating, child sexual abuse, reckless driving). Both agreeableness and conscientiousness tend to increase from early adulthood to middle adulthood, while external-izing decreases.

Ironically, bright people with low agreeableness often make the most revolutionary creative contributions in the arts and sciences, because they want to leave their mark on the world and don't much care what others deem to be conventionally correct. When you think of people whose public personas seem low in agreeableness (hence high in social dominance and tough-minded independence), don't just think of Joseph Stalin or Dick Cheney. Also think of Elizabeth I, Isaac Newton, Frank Lloyd Wright, Ayn Rand, James B. Watson, Margaret Thatcher, Quentin Tarantino, and comedians Larry David (of *Curb Your Enthusiasm*) and Sarah Silverman (of *Jesus Is Magic*). Also, low-agreeableness "bad" boys and girls can be more sexually attractive than "nice" boys and girls, at least for short-term mating, since they are perceived as more assertive, self-confident, exciting, and cocky.

Some agreeableness-advertising bumper stickers:

### High Agreeableness

- Commit random acts of kindness and senseless beauty
- Coexist
- Who would Jesus bomb?
- Live simply that others may simply live
- God bless the whole world. No exceptions
- Has anger solved your other problems?

### Low Agreeableness

- You suck and that's sad
- Sniper: Don't bother running, you'll just die tired
- Just because I'm a mom doesn't mean I care
- If you can read this, you're in my kill zone
- Remember: Pillage, *then* burn

- Yes, this is my truck. No, I won't help you move
- Honk if you've never seen an Uzi fired from a car window
- If at first you don't succeed, blame marketing

(The low-agreeableness category boasts an extraordinary wealth of examples, perhaps because it deters tailgating and preempts road rage.)

**S** is for stability, especially emotional stability. Stability means adaptability, equanimity, maturity, stress resistance. People high in stability are resilient: usually optimistic, calm, at ease, and quick to rebound from setbacks. People low in stability are neurotic: anxious, worried, self-conscious, depressed, pessimistic, quick to anger, quick to cry, and slow to rebound from setbacks. Low stability corresponds with the "internalizing" dimension of mental illness that is associated with distress (major depression, dysthymia, generalized anxiety disorder) and fear (phobias and panic disorder). High stability correlates positively with general mental health and general happiness, including job satisfaction and marital satisfaction. In fact, in the developed world, emotional stability predicts overall life satisfaction more strongly than does income or any of the other Central Six traits.

Some stability-advertising bumper stickers:

## High Stability

- If at first you don't succeed, redefine success
- Not all who wander are lost
- The best things in life are not things
- Smile and let it go
- It's not life's job to make you happy; it's your job to make life happy

## Low Stability

- Can't sleep. Clowns will eat me
- It's not whining if you wave a handgun

- A barrelful of monkeys would not be fun. It would be horrifying
- Tell me, where is this bright side you speak of?
- Some days it's just not worth gnawing through the straps

**E** is for Extraversion: how friendly, gregarious, talkative, funny, expressive, assertive, active, excitement seeking, and socially self-confident one is. Extraverts are social; introverts are loners. Almost all psychologists from Carl Jung onward have agreed that extraversion is a key dimension of individual differences. Shyness arises as a combination of low extraversion and low stability. Extraverts also show higher "surgency"—higher levels of activity, power, dominance, and self-confidence. They show a lot of positive emotions, prefer working with and trusting others, enjoy leadership, and prefer being physically active. They go to parties more and drink more. They are more sexually adventurous and unconventional. Low extraversion is not just associated with shyness; it is also associated with social passivity and low levels of social status seeking. People low in extraversion tend to suppress positive feelings, prefer working alone, prefer being physically passive, and are less trusting and less inclined to seek leadership roles. Since extraverts are more active and meet more people, they tend to have more friends and sexual partners.

Some extraversion-advertising bumper stickers:

## High Extraversion

- If it weren't for physics and law enforcement, I'd be unstoppable
- I'm having so much fun, I could poop
- My cultlike following is now accepting applications
- Strangers' candy is the sweetest
- Adrenaline is my drug of choice

## Low Extraversion

- Hi. Where's your off button?
- If cats could talk, they wouldn't
- You are reinforcing my inherent mistrust of strangers
- Eagles don't flock
- I'm not good at empathy; will you settle for sarcasm?

**GOCASE** (pronounced "go-case") is a silly but useful acronym for remembering these Central Six, conjoining the general-intelligence trait (G) and the Big Five personality traits (O, C, A, S, E). One can think of GOCASE as a social judgment heuristic: when we meet someone new, we can ask ourselves, is this person, a specific case of humanity, a "go-case" with whom we should go ahead and keep talking to find out more, or a "stop-case" whom we should stop interacting with and gently extricate ourselves from? If we meet strangers at a party, and we judge that they are reasonably similar to us on a sufficient number of the Central Six GOCASE traits, it is usually worth talking more with them. The same is true if we are meeting a neighbor, potential employee, first date, or second cousin at a family reunion. In each case, a surplus of intelligence can compensate for a deficit of stability (symptoms of neuroticism); a surplus of agreeableness (warmhearted-ness) can compensate for a deficit of extraversion (reserve or shyness). We don't often meet adults who are very high on the Central Six traits that best predict social status (intelligence, conscientiousness, stability, extraversion), because they typically become so successful so quickly that they rarely interact with ordinary folks like us. (Think Oprah Winfrey, Tony Blair, Ani DiFranco, Elton John, Jodie Foster, Denzel Washington.) Conversely we don't tend to meet adults who are very low on intelligence, openness, conscientiousness, agreeableness, and stability, because they're already dead, or in high-security psychiatric prisons, or protected by Secret Service agents.

Most of these traits are so universally recognized and valued that they even appear in The Sims computer games that simulate consumer behavior. The Sims have five personality dimensions in these games, which are pretty similar to some of the Central Six:

- playful versus serious (which is comparable to high versus low openness)
- neat versus sloppy (high versus low conscientiousness)
- outgoing versus shy (high versus low on the social facet of extraversion)
- active versus lazy (high versus low on the surgency facet of extraversion)
- nice versus grouchy (high versus low agreeableness)

The game designers at Electronic Arts may not have read much personality psychology, but they did intuit which individual differences drive human social interactions and consumption preferences. (They didn't include stability as a Sims trait, because highly stable Sims would be highly happy regardless of their material or social circumstances, so they'd short-circuit the whole point of the game, which is to make one's Sims happy through their income, consumption, and social networking.)

All human cultures seem to have their own terms for these traits, and to value them when selecting mates, friends, and business associates. For example, an individual's reputation in Chinese society has depended for centuries on two key attributes: *mianzi* and *lian*. *Mianzi* concerns other people's perceptions of a person's status, prestige, or "face," which would usually reflect his general intelligence and extraversion. A loss of *mianzi* would entail a loss of authority, respect, and influence. *Lian* concerns other people's perceptions of a person's moral virtues, which would usually reflect his conscientiousness and agreeableness. A loss of *lian* would entail a loss of trust and perceived virtue—which could have terrible consequences when social and business relations are based on moral reputation rather than contract law.

Personality traits exist not only across human cultures, but also across animal species. The biologist David Sinn has shown that a basic bold-versus-shy dimension of temperament can be discerned even among squid. This trait is roughly a combination of extraversion and stability, and it shows a moderately strong genetic heritability in predicting responses to threats from predators. Personality traits seem to progressively differentiate as brains evolve to be larger and social interactions more complex. Samuel Gosling showed that hyenas have several personality dimensions that can be reliably rated: assertiveness (which is similar to low agreeableness), excitability (similar to low stability), human-directed agreeableness (similar to high agreeableness), sociability (similar to high extraversion), and curiosity (similar to high openness). Gosling also showed that domesticated dogs show analogues of four out of the Big Five traits—energy (extraversion), affection (agreeableness), emotional reactivity (low stability), and intelligence (openness/intelligence)—and that humans can rate these traits in dogs just as reliably they can in other humans. The Big Five also seem applicable to five other species of pets (cats, ferrets, rabbits, hedgehogs, horses), and to the other four great ape species (gorillas, orangutans, chimpanzees, bonobos). Thus, the Big Five are likely to have existed for at least 13 million years (back to the last common ancestor of all great apes), and possibly for as long as 125 million years (back to the common ancestor of all mammals). Some personality dimensions may even date from the divergence between vertebrates and invertebrates, around 600 million years ago.

## How Were the Big Five Discovered?

To most people, general intelligence sounds like a genuinely important dimension of human variation, whereas the Big Five personality traits sound a bit faddish and arbitrary. How can we have confidence that these five traits have any more reality and predictive validity than other personality constructs that are now consigned to the ever-fuller dustbin of psychology's history? Aristotle's student Theophrastus (371–287 BC) wrote a book on thirty personality types. Galen of Pergamum

(AD 130–200) described four personality dimensions corresponding to the four humors (sanguine, choleric, phlegmatic, melancholic),
which constituted the dominant European theory of personality
throughout the Middle Ages and the Enlightenment, persisting even
into eighteen-century philosophy (Immanuel Kant) and nineteenth-
century psychology (Wilhelm Wundt).

Twentieth-century personality psychology developed more sophisticated methods and statistics to characterize human individual differences, but for many decades, there was no consensus on the number
or nature of personality dimensions. In the 1940s, Raymond Cattell
proposed sixteen dimensions of personality, whereas the Myers-Briggs
system advocated four dimensions. In the 1970s, Hans Eysenck argued
for three dimensions. Their models have faded. Why should we think
the Big Five will show any more staying power?

The Big Five theory is distinctive because it emerged gradually,
over the course of decades, through cumulative empirical research
and consensus-building literature reviews. The five dimensions are
not the product of one researcher's pet theory. In 1936 Gordon Allport and Henry Odbert compiled 4,500 English adjectives describing
human personality, and later researchers found that meaning similari-
ties among these adjectives could be reduced to about five dimensions.
These same Big Five dimensions were first empirically identified in
1954–1961 by two U.S. Air Force researchers (Ernest C. Tupes and
Raymond E. Christal) by analyzing personality data from eight large
samples of people. Their results were replicated by Warren Norman
in 1963 in another large sample. In 1981 four leading personality
researchers (Andrew Comrey, John Digman, Lewis Goldberg, and
Naomi Takemoto-Chock) reached a consensus that the Big Five were
a reliable way to summarize the empirical personality literature.

Throughout the 1980s, the Big Five model was widely accepted in
psychology as a viable replacement for previous models. Paul Costa and
Robert McCrae found the five factors robustly across virtually all the
personality scales that were being used at the time. They published a
Five Factor personality inventory in 1985, and have been refining such

measures since then. In the 1990s each of the Big Five proved to be highly stable across time within individuals, about 50 percent genetically heritable across generations, and fairly universal across cultures. For example, translated Big Five personality questionnaires still yield five factors when administered to Chinese, German, Hebrew, Korean, Japanese, and Turkish subjects. Also, across cultures, women always show higher average scores than men on agreeableness and conscientiousness, and lower average scores on stability.

Strictly speaking, I've muddled together a couple of different five-factor personality theories. Lewis Goldberg's Big Five model includes five dimensions called intellect (which I call openness), conscientiousness, agreeableness, emotional stability (which I call stability), and surgency (which I call extraversion). By contrast, Costa and McCrae's NEO Five Factor Model includes five similar dimensions called openness, conscientiousness, agreeableness, neuroticism (which I call low stability), and extraversion.

The technical names of the Big Five may sound a bit obscure, but they each have hundreds of synonyms that we use in ordinary conversation. This is the surprising upshot of the 1936 Allport and Odbert adjective list: the Big Five personality traits loom so large in human life that we have invented vast arrays of words for describing them when we gossip about others. Personality psychologists call this the lexical hypothesis: natural-language personality terms already capture fairly accurately the true dimensions of human variation. Indeed, if there was a major dimension of human behavioral variation that had not already been noticed and lexicalized by our linguistic ancestors, it would be surprising indeed.

Language aside, a person's scores on the Big Five yield a lot of useful information about how to interact with him or her. Lewis Goldberg argues that when we meet people, we come armed with a few fundamental questions about them: (1) are they interesting (open) or boring?; (2) are they reliable (conscientious) or flaky?; (3) are they nice (agreeable) or nasty?; (4) are they sane (stable) or crazy?; (5) are they dominant (extraverted) or submissive? In fact, when I take my border collie, Jenny, to the public dog parks of Albuquerque, these are exactly the same questions I ask about any new dog I encounter—especially questions 3 and 4. If I meet a pit bull, I know it is likely to be nasty and crazy, because pit bulls have been selected across generations for dogfighting ferocity. This is why they deserve to be illegal (as they are in Australia, Britain, France, New Zealand, and Miami, Florida), and why the *Onion* newspaper can raise a chuckle with its mock headline "Heroic Pit Bull Journeys 2,000 Miles to Attack Owner" (April 17, 2002). Whereas, if I meet another border collie, I know it is likely to be nice and conscientious, just like its discerning and virtuous owner.

## Preferences for the Central Six

There is a key difference to be noted between intelligence and the Big Five. For intelligence, most people prefer friends, mates, and allies who are brighter than average rather than dumber than average. Folks may not want an IQ-160 genius to occupy every niche in their social network, but they do generally wish to have an IQ-120 pal rather than an IQ-80 pal, whatever their own IQ level. Within the normal range of intelligence, higher is generally better, and is more valued socially and sexually. So, G has a strong directionality: everybody wants it, and everybody wants to display it. Only when we are trying to take advantage of others do we prefer to interact with lower-intelligence people—when we con them, seduce them, sell them things, steal their resources, or fight wars against them.

However, for the Big Five personality traits, preferences are more variable. Generally, people like others whose own personality traits are pretty similar to their own. Those who are high-openness prefer to date, marry, befriend, and work with high-openness others. Then they can gossip happily about avant-garde science, culture, and aesthetics— who's won recent Nobel, Booker, Turner, or MacArthur prizes. Those who are low-openness equally prefer low-openness others, so they can gossip happily about how the pretentious, artsy, metrosexual know-it-alls are undermining religion, tradition, and civilization itself. Likewise, the highly agreeable and gentle like to hang out in churches, nonprofit groups, co-ops, and nonviolent peace demonstrations with other highly agreeable people; whereas the highly disagreeable and assertive prefer to congregate in war zones, organized crime, and lobbyist conferences. The highly extraverted can be found at parties flirting with one another, while the introverted stay home, or sit in libraries a few tables away from one another, reading quietly.

One's self-esteem usually tracks the traits that are socially valued by others. High intelligence and high physical attractiveness tend to increase self-esteem, whereas the opposite traits lower it, because others usually value high intelligence and attractiveness. Likewise, if one extreme of a Big Five trait was consistently preferred across most situations by most people, then we might expect that people at that extreme would develop much higher self-esteem than those at the other extreme. The psychologist Richard Robins and his colleagues did a huge online study of 326,641 people. They found that self-esteem correlated positively with openness ($r = .17$), conscientiousness ($r = .24$), agreeableness ($r = .13$), stability ($r = .50$), and extraversion ($r = .38$). Nine previous studies had also found that self-esteem was substantially predicted by emotional stability (happiness, freedom from anxiety, worry, and depression), and modestly by conscientiousness and extraversion, but not consistently by openness or agreeableness. These ten studies suggest that people usually value higher levels of stability, conscientiousness, and extraversion in others, but that preferences for openness and agreeableness are not consistent. Further studies have shown that people consider higher levels of conscientiousness, stability, and agreeableness to be more "normal" than lower levels, whereas lower levels of openness were considered more "normal." Altogether, such studies suggest that there are modest social preferences for others to show higher rather than lower levels of conscientiousness, agreeableness, stability, and extraversion.

For certain domains of social choice, people do have directional preferences for higher- or lower-than-average levels of particular Big Five traits. Managers of microchip factories usually want workers who are highly conscientious, agreeable, and stable, while managers of creative advertising teams may prefer workers who are highly open and extraverted. Our preferences for Big Five traits can reverse, adaptively and sensibly, depending on the circumstances. High impulsivity and spontaneity may be attractive in a short-term lover, but high conscientiousness may be preferred for a long-term spouse who is responsible for child care and mortgage bills. We may have mixed feelings about an acquaintance's Big Five traits, since each level of each trait entails both costs and benefits. Highly extraverted friends make it easier to make even more friends, but they're also more likely to seduce your spouse. Highly agreeable co-workers may be more pleasant to work with on most days, but they may be too wimpy to join a strike for better safety conditions.

Much of human social intelligence seems dedicated to discerning which kinds of people, with which Big Five personality traits, would be most useful at any given moment, given the particular challenges we face. Need a couples therapist? Choose someone high on agreeableness. Need a bodyguard? Choose the opposite. Need a tax accountant? High conscientiousness would work better. Need a friend for a wild bachelor party in Las Vegas? Low conscientiousness would party harder.

Equally, we have incentives to bias how we present ourselves to others, depending on their needs and circumstances. The couples therapist needs to act ferociously disagreeable if she's in danger of being mugged. The tax accountant needs to act like a frolicsome, low-conscientiousness free spirit if her husband complains that she is too dull in bed. This is the essence of "impression management," and it is a central social skill that humans acquire throughout childhood and adolescence. We learn to present our apparent Big Five traits in adaptively biased ways. Normal adults learn this skill so well that we recalibrate our trait displays dozens of times a day, to suit our audience, goals, and environments. (People who are high on a personality trait called self-monitoring are especially prone to monitor and shift their displayed personality as a function of their social environment.) The

Big Five traits are stable across each individual's life span not because they predict behavior invariably across all situations, but because they predict behavior on average if you get to know somebody across many different situations.

The most dramatic shifts in apparent personality are called emotions. If we suddenly need to appear much lower in agreeableness, given some social threat, we enter a special new mode of perceptual, cognitive, and behavioral operation called anger. Anger gives us a credible but temporary boost in assertiveness, aggressiveness, and formidability. If we suddenly need to appear much higher in openness in order to attract a particular mate, we enter a special mode of operation called being in love. This emotional state yields a huge but temporary boost in energy, novelty seeking, and interest in culture, poetry, music, the arts, the emotional nuances of human interaction, and the existential mysteries of the cosmos. Once the courtship display accomplishes its mission (which might range from copulation to marriage), the being-in-love dissipates, and we return to our normal, usually much lower level of openness. Less dramatic shifts in apparent personality are called moods—they last longer than emotions, but are less extreme in intensity. An irritable mood reduces one's agreeableness; a whimsical mood reduces one's conscientiousness.

So, there is continuum of duration from emotions and moods as short-term states to Big Five personality dimensions as long-term traits. From the perspective of emotions research, personality traits are simply stable propensities to feel certain emotional states more often. But from the perspective of personality research, emotions are simply transient shifts in one's manifest personality traits.

This duration difference between emotional states and personality traits has obvious implications for the ways we evaluate others. If we are seeking a one-shot, short-term interaction with someone, we pay more attention to the person's current emotional state. If we just want to buy a few groceries in a store that we will never enter again, we may choose the cashier who seems happier (apparently more agreeable) at the moment, without bothering to judge whether that agreeableness level will be stable across time. But if we are seeking a long-term relationship with someone, we pay more attention to his or her stable personality traits. If we frequent the local organic food co-op, we may come to favor the cashier who is consistently agreeable, even if he has some bad days. Since the most important relationships in life tend to be such longer-term, repeated interactions, we care more about judging people's stable personalities in those contexts.

## Measuring Your Big Five

The Big Five can now be measured with moderate accuracy, using a self-rating scale, in about one minute. The psychologists Beatrice Rammstein and Oliver John published a Big Five scale in 2007 called the BFI-10 that uses just ten questions. They found that people's scores on this very short scale were very reliable across a two-month period (with test-retest correlations about .84), and correlated very highly (about .82) with their scores on much longer personality scales.

Their BFI-10 scale is reprinted below in a slightly clearer form; try it and see how you score.

After each statement below, write down a number from 1 to 5 to represent how well the statement describes your personality, where

>   1 = disagree strongly,
>   2 = disagree a little,
>   3 = neither agree nor disagree,
>   4 = agree a little,
>   5 = agree strongly.

### I See Myself as Someone Who

1. has an active imagination. _____
2. has few artistic interests. _____
3. does a thorough job. _____
4. tends to be lazy. _____

5. is generally trusting.                    _____
6. tends to find fault with others.     _____
7. is relaxed, handles stress well.     _____
8. gets nervous easily.                    _____
9. is outgoing, sociable.                  _____
10. is reserved.                              _____

Here's how to score yourself. Items 1 and 2 concern openness; 3 and 4 concern conscientiousness, 5 and 6 concern agreeableness, 7 and 8 concern emotional stability, and 9 and 10 concern extraversion. For each successive pair of items, subtract the number you wrote for the even-numbered item from the number you wrote for the odd-numbered item, and that gives your score for the corresponding Big Five trait. Scores can range from −4 (very low on the trait) to +4 (very high on the trait), with 0 being about average.

For example, if you "agree a little" with "I see myself as someone who has an active imagination," you should have written a 4 for item 1. If you "disagree strongly" with "I see myself as someone who has few artistic interests," you should have written a 1 for item 2. Then you'd subtract your response to the even-numbered item (1) from your response to the odd-numbered item (4), yielding a score of 3—which would mean you are quite high on openness, given that the average is 0 and the maximum is 4.

## The Central Six Each Form a Bell Curve

Everybody knows that the distribution of human intelligence forms a bell curve, or a roughly normal distribution. Most people cluster in the middle, near IQ 100. The distribution tapers off quickly as IQ scores deviate from the average, so that fools and geniuses are both rare. When most of us seek mates, friends, or co-workers, we are not much bothered about the extremes, but do the best we can to distinguish among others near the middle of the range. We generally prefer associates with IQs of 115 rather than 95 because that stacks the odds in our favor: they are a bit more likely to help us solve existing problems, and a bit less likely to create new ones.

Less well-known is the fact that all the Big Five personality traits have a similar bell-curve distribution. Most people are moderately agreeable—capable of warmhearted kindness under some circumstances, capable of wicked selfishness under others, but generally just muddling along in the seminice, semiselfish state we recognize as the human condition. Relentless good and relentless evil are equally rare.

This bell-curviness of each of the Central Six dimensions is a genuine empirical discovery. While it is true that almost all continuously varying biological traits form a bell curve, many of these traits are not continuously varying; they are discrete. They involve distinctive patterns of gene activation that give rise to qualitatively different things: neurons versus muscle cells versus bone cells. Brains versus hearts versus femurs. Males versus females. Sexually immature juveniles versus sexually mature adults. Caterpillars versus butterflies. Fertile versus pregnant females. It is fairly common for species to have distinctive "morphs" of this sort: different forms or states of the organism specialized for different social, sexual, or ecological roles. Species themselves are, of course, different morphs on a grander scale, kept qualitatively distinct by reproductive isolation.

So, it could have been the case that human personalities fell into discrete categories, like Jungian archetypes: the child, the eternal boy, the hero, the great mother, the wise old man, the trickster. However, most of these are just typical human life stages rather than distinctive personality types: child = child, eternal boy = narcissistic adolescent; hero = single young male pursuing reproductive success through high-risk status seeking; great mother = mature female; wise old man = mature male. The trickster is a true personality exemplar, but he basically just shows low conscientiousness by breaking rules and violating social norms, plus higher than average intelligence and openness, and lower than average agreeableness. (Note that the marketer stereotype closely follows the trickster archetype.) With regard to all the other interesting personality traits, there aren't salient Jungian archetypes or discrete personality categories—just the continuous normal distribution of each trait.

This simple fact has profound implications for marketing, as it means that most distinct personality types used in market segmentation are illusory. The outdated Myers-Briggs dichotomies (feeling versus thinking; judging versus perceiving) just can't work if the underlying traits are normally distributed. Also, if many of the demographic (age, sex, ethnicity) differences in consumer behavior boil down to these different groups' having somewhat different average scores on the Central Six dimensions, then these demographic categories will also be deeply misleading in characterizing consumers. For example, if men respond more positively to an aggressive-looking product design than women do, it may be tempting for market researchers to attribute the difference to sex per se. However, males and females have different average levels of agreeableness, which probably influence reactions to aggressive-looking products. The sex difference in agreeableness, and not sex itself, may be the factor driving the consumer responses. It is important to know which is the case, because agreeableness is continuous, whereas sex is dichotomous. The distribution of male agreeableness overlaps substantially with the distribution of female agreeableness, so measuring consumer agreeableness levels may be much more predictive of their reactions than asking about their sex. The same concern applies to all other criteria for market segmentation. Nation, region, language, culture, socioeconomic status, class, and education level may predict consumer behavior mainly because they are correlated with some of the Central Six traits, not because they directly cause the behavior. If so, it will almost always be more effective to measure the Central Six directly rather than relying on traditional market segmentation categories to predict behavior.

## The Central Six Are Fairly Independent

Surprisingly, the Central Six traits are not much correlated with one another. In fact, they are almost statistically independent: knowing a person's score on one trait gives you almost no information about their other traits. This explains why people are motivated to display all six traits in different ways, through different behaviors and product purchases.

The only major exception is that general intelligence has a modest positive correlation with openness: bright people tend to be more interested than average in new experiences, travel, culture, and aesthetics. Conversely, people who are culturally engaged and open-minded tend to be brighter than average. For this reason, university towns tend to have better cultural institutions. Yet even here the small size of the positive correlation means that there are plenty of bright but conventional people, who may work as engineers for the military-industrial complex, listen to the same classic rock they did thirty years ago, and know forty gigabytes of baseball statistics. Likewise, there are plenty of open-minded novelty seekers who love strange ideas and experiences, but who are not very bright. They constitute the market for fantasy novels, self-help books, nutraceuticals, facial piercings, music by Enya, degrees in nonevolutionary psychology, and every product labeled "homeopathic." Indeed, their combination of neophilia and inanity make them an extremely profitable market segment.

The fact that the Central Six are fairly independent of one another violates many of our social stereotypes. You might know some graduate students and conclude that high intelligence and high neuroticism (low stability) go hand in hand. Yet they don't, on average, across the whole population: lower-intelligence people get just as anxious and depressed. You might know some Greenpeace liberals and assume that high openness and high agreeableness go together. Yet they don't: some people seek extreme new experiences in sadomasochistic unsafe sex, criminal gangs, political terrorism, or careers in cosmetics marketing (think of Tyler Durden in *Fight Club*). For every pair of social stereotypes that seem to confirm a positive correlation between two of the Central Six, there exists an opposite pair of stereotypes that seem to confirm a negative correlation.

Our stereotyped dichotomies are so numerous and nuanced that almost any combination of personality traits can seem to fit some type, yet those types are illusory. As we have seen, in the space of human personality traits, we really have what statisticians call a multivariate normal distribution: each dimension is a bell curve with most people near the average, and each dimension is independent of the others. Given our six independent dimensions, if we split each into just three levels (low, average, or high), then we'd have three to the sixth power possible combinations, or 729 different personality types—rather larger than the number of types typically posited in astrology, Jungian psychoanalysis, or most market segmentation.

## Beyond the Central Six?

Clearly, the Central Six do not cover all the individual differences that characterize human nature, for we must also take into account virtues and vices, values and interests, political and religious attitudes, hobbies and skills, mental illnesses and addictions. Yet even many of these can be predicted fairly strongly by the Central Six. They correlate, often to a surprisingly high degree.

For example, many of the new-fangled types of intelligence that have become popular recently (social intelligence, emotional intelligence, creative intelligence) boil down to general intelligence plus some combination of the Big Five personality traits. Social intelligence as studied by developmental psychologists and primatologists means the capacities for perspective taking and social strategizing, but it seems rather well predicted by a combination of general intelligence and extraversion, plus agreeableness (when empathy pays) or disagreeableness (when exploitation pays). People with autism have only somewhat reduced average intelligence, but they typically have severely reduced extraversion and agreeableness. Similarly, emotional intelligence means the capacities for perceiving emotions expressed by others, using emotions to guide one's own thinking and problem solving, understanding the nature and social functions of emotions, and managing one's own emotions adaptively. Yet the capacities for perceiving and understanding emotions correlate strongly with general intelligence, and emotional self-management correlates strongly with conscientiousness and stability. Finally, creativity research suggests that short-term creative intelligence is basically general intelligence plus openness, while long-term creative achievement is also predicted by conscientiousness (hard work and ambition) and extraversion (active surgency and social networking). In our recent book *Mating Intelligence*, Glenn Geher and I posited a dimension of "mating intelligence" concerned with sexual courtship and relationships—but unlike the advocates of social, emotional, or multiple intelligences, we explicitly argued that it is likely to correlate rather strongly with general intelligence and some of the Big Five.

Sexual traits are also well predicted by the Central Six. The personality trait of "sociosexuality," as developed by Steve Gangestad and Jeffry Simpson, indexes sexual promiscuity. People with "unrestricted" sociosexuality (the highly promiscuous) have larger numbers of sexual partners, more one-night stands, and higher infidelity rates. They also tend to have high extraversion. People who are very extreme on these traits tend to join the Lifestyle—the polyamorous community of hooking up, swinging, and open marriages exemplified by the International Lifestyle Association. The highly sociosexual, open, impulsive, and selfish tend to invest more of their time and energy in "mating effort" rather than "parenting effort": they are constantly seeking new sexual partners rather than raising the offspring from existing relationships. On the other hand, people with "restricted" sociosexuality (the virginal, the chaste, and the happily married) have fewer sexual partners, less infidelity, lower openness, higher conscientiousness, higher agreeableness, and lower extraversion. They invest more time and energy in parenting effort and less in mating effort. The highly sociosexual consider the less sociosexual to be repressed and sanctimonious, and dismiss them as prigs, prudes, puritans, and hypocrites.

The less sociosexual denounce the more sociosexual as sluts, tarts, whores, rakes, and dogs. (Clearly, there are different opinions about the optimal level of sociosexuality that one's mates, friends, and neighbors should display.)

The Central Six also predict social, political, and religious attitudes fairly well. Liberals are only a little brighter than conservatives on average, but they tend to show significantly higher openness (more interest in novelty and diversity), lower conscientiousness (less adherence to conventional social norms), and higher agreeableness (more widespread empathy and "bleeding hearts"). Conservatives show lower openness (more traditionalism and xenophobia), higher conscientiousness (family-values moralism, sense of duty, civic-mindedness), and lower agreeableness (more hard-headed, hard-hearted support for their self-interests and national interests). However, since the traditional left-right political spectrum has only one dimension, and the Central Six has six dimensions, it is more accurate to describe the complete range of human political attitudes using the Central Six. For example, the 1960s New Left was basically more open (freethinking) than the 1930s Old Left. Fascists can be seen as basically lower-intelligence conservatives with lower stability (more fear, distress, anxiety, and neuroticism) and even lower agreeableness (more aggressive interests in warfare, torture, and genocide). Libertarians can be viewed as basically higher-intelligence liberals with slightly higher conscientiousness (faith in social reciprocity and the work ethic), lower agreeableness (distaste for conspicuous sympathy displays), and an extra dollop of extraversion (self-reliant surgency).

Although the Central Six are fairly solid empirical discoveries, research on individual differences continues, and might hold some surprises. General intelligence is unlikely to be dethroned as the queen of predictive power in the land of psychodiversity, but the Big Five might be replaced with an even better model of personality variation at some future point. This will depend on new discoveries in genetics (do the Big Five depend on distinct sets of genes?), neuroscience (do the Big

Five depend on distinct brain systems?), and evolutionary psychology (do the Big Five serve distinct adaptive functions?). We need a much clearer idea of why evolution should have maintained heritable variation in five main personality traits, rather than three, or eight, or fifty. The answer must have something to do with the number of ways that human social strategies could vary adaptively within prehistoric clans. Several researchers are working hard on this issue in the new field of evolutionary personality psychology. But for the moment, the Big Five is the best model we have, and we might as well see how far we can go with it.

# 10

# Traits That Consumers Flaunt and Marketers Ignore

IF WE WERE all honest about these Central Six traits, then dating, socializing, and working would be much simpler. We could just have six numbers tattooed on our foreheads, representing our percentile scores on each of the traits. Speed dating could be even faster, as we could instantly reject potential mates who score too low or too high on the traits we care most about. Presidential debates could just be one-minute close-ups of each candidate's trait percentiles, and we could immediately see which candidate is too stupid (or too smart?), or too high or low on each of the Big Five, for us to vote for him or her.

Unfortunately, such trait-score tattoos would be unreliable, as people would be naturally inclined to fake them. Consider the unreliability of bumper stickers as trait displays. Because our mental traits are hard to assess when we're wrapped in sheet metal and tinted glass, as we saw earlier, drivers often decorate their cars with $4 trait displays called bumper stickers, in an attempt to reveal the driver's soul on a three-by-ten-inch plastic-laminate strip. However, there is no way to guarantee that bumper stickers accurately reflect the driver's true mental traits. Convenience-store clerks will sell you these things without even checking your scores on the one-minute BFI-10 personality inventory. You might have even shoplifted one of the high-conscientiousness or high-agreeableness examples. Economists have a special technical term for a signal that is so unreliable—they call it "cheap talk." Bumper stickers are promises with no credibility, claims with no evidence. They may be amusing, but you would not choose a friend or lover solely on that basis.

So, instead of just displaying cheap-talk trait tattoos and bumper stickers, we buy and display costly products that we think will testify more reliably to our key traits. Many people spend tens of thousands of dollars and four years of hard study getting a prestigious university degree, which contains no more trait information than a two-hour IQ test. We pay huge interest on credit-card debt just to build up a good credit score, to credibly display our conscientiousness for the time when we need a mortgage. We spend profligately on mobile phones so we can gossip loudly in public to our friends and appear credibly extraverted. Some of us subscribe to *Harper's, Wired,* or *Prospect* to display cosmopolitan cultural sophistication, also known as openness. Billions of others pray daily, attend church weekly, and support the costly priest caste to display their moral virtues—conscientiousness, agreeableness, conservatism.

What these products have in common is not just their up-front capital cost, but the difficulty of exploiting them properly if one lacks the right personality traits. Even if you do manage to fake your way into Cambridge University, you won't be able to fake your essays, tutorials, class participation, and grades. Even if your income qualifies you for an American Express Platinum Card, your credit rating will drop if you don't pay your bills conscientiously. Even if you subscribe to *Prospect* (Britain's best intellectual magazine), you won't be able to converse intelligently about its content if you are not open to new ideas. If you are an irritable, short-tempered psychopath, you will not be able to stand weekly church sermons and rituals that will test your limited patience and agreeableness to the breaking point.

Thus, some of the most socially important products cannot be bought and displayed with money alone, because we want to know a lot more about people than their wealth. Many of these products don't even look like products in the traditional retail sense. A Cambridge degree, a good credit rating, a local reputation as a well-informed intellectual or a generous churchgoer—none of these can be bought in a shopping mall, although money is required for each. In the next four chapters, we'll look at the four traits out of the Central Six that I consider most important for understanding consumer behavior: intelligence, openness, conscientiousness, and agreeableness. Stability and extraversion are also fascinating, but it would be tedious to analyze all the Central Six in equal detail. By the time you understand the first four in the coming chapters, you'll probably get the idea.

## How Car Choices Reveal the Central Six Traits

One of the most expensive ways that consumers try to display their Central Six traits is through their choice of car brands and features. At least at a semiconscious level, car buyers seek a match between their own personality traits and the apparent "brand personality" promoted in the carmaker's advertising. They also tend to seek the features that seem most important given their aspirations and anxieties as shaped by their Central Six traits. The table below lists some car brands and features that seem associated with high and low levels of each of the Central Six. (These are just my own impressions and stereotypes; each entry is debatable and subject to many exceptions. As with most products, there seems to be no good data on which brand personalities or car features are actually associated with each of the Central Six.)

### High Intelligence
- Favorite brands: Acura, Audi, BMW, Lexus, Infiniti, Smart Car, Subaru, Volkswagen
- Favorite features: maximum value, complex controls, reading lights, headroom, hard-to-pronounce brand name and/or model name

### Low Intelligence
- Favorite brands: Cadillac, Chrysler, Dodge, Ford, GMC, Hummer
- Favorite features: large mass, low down payment, dealer financing, high size-to-reliability ratio

### High Openness (*Liberalism, Eccentricity*)
- Favorite brands: Lotus, Mini, Scion, Subaru
- Favorite features: eccentric design, foreign origin, ground clearance, moonroof, popularity among youth

### High Conscientiousness (*Responsibility, Caution*)
- Favorite brands: Acura, Honda, Lexus, Volvo, Toyota
- Favorite features: reliability, child-safety locks, antitheft alarm, daytime running lights, gas mileage

### Low Openness (*Traditionalism, Conservatism*)
- Favorite brands: Buick, Lincoln, Oldsmobile, Range Rover, Rolls-Royce
- Favorite features: traditional design, domestic origin, popularity among the elderly and royalty

### Low Conscientiousness (*Impulsiveness, Recklessness*)
- Favorite brands: Ferrari, Jeep, Mitsubishi, Pontiac
- Favorite features: cruise control, cup holders, high acceleration

### High Agreeableness (*Kindness, Gentleness, Altruism*)
- Favorite brands: Acura, Daewoo, Geo, Kia, Saturn
- Favorite features: eco-friendly design, hybrid drive, payload to help friends move, smiley-looking front end

### Low Agreeableness (*High Aggressiveness, Dominance*)
- Favorite brands: BMW, Hummer, Maserati, Mercedes, Nissan
- Favorite features: horsepower, torque, intimidating size, menacing design, leather seats, sneering front end

### High Stability (*High Happiness, Self-Esteem*)
- Favorite brands: Acura, Porsche, Scion
- Favorite features: cheerful design, happy "vibes"

### Low Stability (*High Anxiety, Neuroticism, Worries*)
- Favorite brands: Volkswagen, Volvo
- Favorite features: safety, airbags, antilock brakes, electronic stability control, extended warranty

### High Extraversion
- Favorite brands: Aston Martin, BMW, Ferrari, Mini, Porsche
- Favorite features: convertible, high-wattage subwoofers, vanity plates, ski rack, product placement in James Bond movies

### Low Extraversion
- Favorite brands: Acura, Hyundai, Lexus, Saab, Subaru, Volvo
- Favorite features: tinted windows, neutral paintwork, quiet interior

Despite the lack of quantitative evidence, some brands clearly strive to become associated with a certain Central Six trait. For example, Subaru sponsors the annual meeting of the American Academy for the Advancement of Science, so it pretty clearly seeks patronage from high-intelligence consumers.

Some of the brands seem associated with extremes of several traits—BMW, for example, seems to connote high intelligence, low agreeableness, high stability, and high extraversion. If only one-third of people are at these extremes for each of these four traits, and if the four traits are uncorrelated, then BMW is tacitly restricting its market segment to one-third to the fourth power, or one out of every eighty-one potential car buyers. Thus, a strong brand personality can allow the consumer to display a more distinctive trait signal, but it might limit a company's market share. By contrast, if Oldsmobile is associated mainly with one trait's extreme (being older than average, hence lower in openness), then it is restricted to only one out of three potential car-buyers.

## Advertising the Central Six Through Music Preferences and Web Pages

In a few exceptional cases, product choices can function similarly to cheap-talk trait tattoos and bumper stickers. Music preferences, for example, seem to work as fairly reliable indicators of the Central Six traits. In some fascinating recent studies, personality psychologists Peter Rentfrow and Samuel Gosling have investigated how people's Big Five traits are conveyed commonly, quickly, easily, and accurately by their stated music preferences. Previous research had already shown that personality traits and music preferences are genuinely correlated, and that people display their music preferences (in conversation, on personal websites, or on iPod playlists) to display their personality traits. Rentfrow and Gosling went further in analyzing the rich personality information conveyed by musical tastes. In one study, sixty college students were asked to get to know one another during a six-week period through an online chat system, and the researchers recorded the topics of their conversations. They found that music was by far the most common chat topic—more popular than discussing movies, books, TV shows, clothing, or sports.

In another Rentfrow and Gosling study, seventy-four college students completed a Big Five personality questionnaire and then listed their top-ten favorite songs. For each student, those songs were then recorded onto a CD that was heard by eight listeners, who judged each of that student's Big Five traits. Correlations between listener judgments and student self-reports were significant for four of the Big Five traits: openness (+.47), extraversion (+.27), emotional stability (+.23), and agreeableness (+.21). (These correlations sound low, but they are impressive given that the listeners had no other information about the students they were judging: no photos or video, or information about age, sex, or race.) This accuracy was mediated by both the music genres preferred (emotionally stable students liked country music, at least if they were from Texas) and the specific acoustic features of the songs (extraverts liked music rated as containing a lot of energy, enthusiasm, and singing). Moreover, the personality information conveyed by music preferences nicely complemented that which can be discerned from photos or short video recordings of people (which are better at revealing conscientiousness and extraversion).

It also seems likely that music preferences reveal general intelligence. Conventional Top 40 radio stations, pop music, and easy-listening music are designed to maximize sales by appealing to the center of the bell curve. Alternative music and classical music basically connote higher-intelligence music—"difficult listening" music—which appeals to a smaller but more discerning market segment. It tends to be more complex with regard to melodic structure and scale, timbral richness and variety, rhythmic intricacy and variety, and lyrical vocabulary and allusiveness. This musical complexity requires more from the listener's auditory perception, attention, and short-term memory, so listeners of lower intelligence find it overwhelming, stressful, and weird. So, higher intelligence can be displayed with some reliability through a stated preference for music by Bartók and Björk, rather than Lynyrd Skynyrd and Hannah Montana.

Thus, a MySpace profile containing nothing more than a face photo and a top-ten list of favorite songs allows reasonably accurate assessment of the Central Six traits, without the individual needing to buy or display any other products. A sneering goth girl who likes Rage Against the Machine, Nine Inch Nails, and Marilyn Manson probably has low conscientiousness, low agreeableness, and low emotional stability. A smirking, well-groomed boy who likes Jars of Clay, Mercy Me, and Third Day (Christian rock groups) probably has the opposite traits. (Lock them together in a Winnebago and you've got an amusing reality TV show.)

How can music preferences as stated on social websites work reliably as trait displays, if trait tattoos and bumper stickers can't? The key to their reliability is that others can easily call your bluff, by interrogating you about your allegedly favorite bands through e-mail and instant messaging. Suppose you claim to love Björk and start corresponding with other Björk fans on Facebook. If you don't really know about her and her music, or have credible emotional and aesthetic responses to its godlike quirky-cosmic genius, other Björk fans will quickly discover your mendacity. You'll lose all credibility if you can't spell her last name (Guðmundsdóttir) or opine that *Vespertine* was her most danceable CD. As Central Six displays, music preferences stated on one's Facebook site are much more reliable than bumper stickers glued to one's car, because it is easier for others to check your bona fides by e-mailing your Facebook site than by getting you to pull over on a highway and talk about your beliefs, desires, and views concerning Icelandic vocalists. Further studies by psychologists Simine Vazire, Samuel Gosling, and others have confirmed that people can judge someone's personality surprisingly accurately by looking at the content of his or her Web page.

## Why Marketers Ignore the Central Six

Surprisingly, most marketers have no idea how well the Central Six can predict consumer behavior. The typical consumer behavior textbook includes a large section devoted to individual differences, but no discussion of general intelligence and the Big Five traits. Rather, the focus is on diverse "factors" that may influence consumer decision making: wealth, time, knowledge, attitudes, values, self-concepts, and motivations. The fact that the Central Six efficiently predict individual variation across all these factors remains unknown or ignored. General intelligence sometimes makes an appearance under the guise of "cognitive resources," but this term is usually construed to mean individual differences in the ability to focus attention given current distractions. The fact that intelligence predicts wealth and knowledge is ignored, as is the fact that the Big Five traits predict attitudes, values, self-concepts, and motivations. Marketers likewise pay attention to "demographic variables"—age, sex, ethnicity, socioeconomic status—without taking into account their correlations with the Central Six.

A similar ignorance of the Central Six afflicts academic consumer-behavior research. The three leading journals—*Journal of Marketing* (JM), *Journal of Marketing Research* (JMR), and *Journal of Consumer Research* (JCR)—have published sixty-four hundred papers in total, of which only about sixty papers refer to personality at all, and only three of which refer to five-factor models of personality or to any of the specific Big Five traits. Individual differences in consumer intelligence get even less attention. Not a single JM, JMR, or JCR paper has ever mentioned general intelligence, the *g* factor, or IQ (one JCR paper in 1984 did mention "cognitive ability," but rejected the idea of a general intelligence factor). References to traits, genetics, and heritability are equally rare. A handful of papers in these journals use costly signaling theory in discussing how companies can send signals about product quality to consumers, but none discuss how consumers can send signals about their own traits to other people. Further, JM, JMR, and JCR papers almost never mention conspicuous consumption, Veblen, social status, or positional goods (discussed below).

In a 2002 review paper, marketing professor Hans Baumgartner observed that:

> Personality research has long been a fringe player in the study of consumer behavior. Little research is directly devoted to personality issues, and if consumer personality is investigated at all, it tends to be from the narrow perspective of developing yet another individual difference measure in an already crowded field of personality scales or considering the moderating effects of a given trait on some relationship of interest.

Basically, consumer research just isn't keeping up to date with discoveries in personality psychology or recognizing how powerfully the Big Five traits influence consumer behavior and self-display.

Many consumer research papers assume that consumers simply prefer products that match their own "identities" or "self-schemata," such that the "brand personality" is congruent with their "consumer personality." The consumer's desire for a product to fit with his personal identity is usually construed as a matter of "relationship quality" (the consumer relates to the brand as if it were a person in its own right) rather than a matter of strategic trait signaling (the consumer chooses the brand to reveal information about his own traits to other real people). By this account, highly agreeable consumers prefer highly agreeable products (those that appear gentle and kind) as a silly side effect of their social preferences for highly agreeable friends—not as a sensible, reliable way to display their agreeableness to others.

Some consumer researchers have recognized that consumers strive to "communicate desired identities" to others through their product choices, especially when the "identity salience" (trait-signaling power) of products is high. This insight is central to "consumer culture theory," which emphasizes the social, cultural, symbolic, aesthetic, and ritualistic ways that consumers create and display their personal and collective identities through their product choices. Consumer culture theory usually entails qualitative observational research on consumer identities within particular markets, subcultures, and ideologies. However, this theory has some serious shortcomings from a scientific point of view: it is vague about the nature of individual consumer "identity" (the specific traits being displayed), murky about the actual function of displaying collective "identity" (class, age, sex, ethnicity), and hostile to evolutionary insights into human nature and social interaction. Also, consumer culture theory has not succeeded in finding common ground with the other main paradigms in consumer research: rational choice theory, cognitive psychology, experimental design, and quantitative analysis.

Part of the problem here is the artificial distinction between "utilitarian" goods (allegedly valued just for their practical utility, such as an axe), "hedonic" goods (valued just for their subjective pleasure, such as ice cream), and "positional" goods (valued just for their status-signaling power, such as diamond earrings). This distinction is treacherous, because, as we saw earlier, all branded products that are profitable to sell must include some elements of conspicuous waste, precision, and/or reputation. These elements inform the consumer's knowledge that the product makes a good trait display, and become confounded with the consumer's experience of pleasure in buying, using, and displaying the product. Also, the ostensible utility of a product is often conflated with its conspicuous precision and reputation. The concept of a positional good is fairly applicable to products as intelligence signals (since most people want to appear positioned higher rather than lower on the bell curve of IQ), but it applies less comfortably to products as signals of Big Five traits (where there is less consensus about which trait level is optimal).

So, most current research on marketing and consumer behavior relies on a chaotic grab bag of outdated theories and unreliable findings. The potent effects of general intelligence are hidden behind its causal effects, empirical correlates, and politically correct euphemisms: education, class, socioeconomic status, consumer knowledge, "cognitive resources," and "cultural capital." Often, marketers think they are studying the effects of class, race, or religion on consumer decision making when they are actually studying the effects of intelligence, which shows different average scores, for whatever reasons, across different classes, races, and religions. The potent effects of the Big Five are likewise hidden behind their correlates and euphemisms: attitudes, motivations, self-concept, values, lifestyle, culture. Here, marketers think they are studying the effects of sex, age, or political beliefs, when they are actually studying the effects of openness, conscientiousness, agreeableness, stability, or extraversion, which also show different average scores across males and females, young and old, liberals and conservatives.

Such imprecise ways of trying to understand individual differences in consumer behavior are popular among marketing professors and consultants for two key reasons. First, acknowledging the Central Six would be politically incorrect, socially awkward, and downright embarrassing for marketers, who mostly think of themselves as the most liberal, progressive, creative people in the corporate world. Marketers are nevertheless surprisingly comfortable with stereotyping people by demographic group (age, sex, ethnicity, class, nationality), despite their political correctness. For instance, one recent consumer-behavior textbook claimed that "Germans are very ambitious, success oriented, and competitive"; "the French seek novelty and elegance"; "Chinese tend to be ethnocentric"; "in Mexico, deadlines are flexible"; "compared with Protestants and Catholics, Jews are more liberal and democratic, more flexible and rationalistic, higher in achievement motivation, more enthusiastic, gregarious, and emotional, more impatient and hurried, more inclined to postpone gratification, and politically most liberal." Marketing is perhaps the last academic field in which blatant stereotyping by nation, religion, class, and sex is countenanced, as marketers can always claim that such groups just happen to have learned different values, norms, and cultures.

It would be much more awkward, socially and ideologically, for marketers to argue that demographic groups have different consumer preferences and behaviors because they have different distributions of psychological traits. Imagine if differences between Asian American and Anglo American consumer behavior could be explained entirely by the slightly higher average IQ of Asian Americans. Imagine if "gender role differences" in consumer behavior boiled down to sex differences in agreeableness. Imagine if some "cross-cultural differences" in consumer behavior boiled down to subtle cross-national differences in average openness or conscientiousness. If marketers simultaneously measured the Central Six traits and their usual demographic/cultural variables, and consistently found differences in trait averages and variances as a function of age, sex, race, class, religion, or nationality, it would be a public relations disaster. Such findings would violate the central intellectual taboo of modern American life—that one mustn't talk about meaningful psychological differences between groups. If, even worse, they found that the Central Six predicted consumer behavior better than the usual demographic/cultural variables, it would cast doubt on the standard liberal blank-slate view that demography is destiny—that socialized identity as part of a collective group predicts human behavior better than individual heritable traits.

Also, group stereotyping (a.k.a. market segmentation) would become more challenging if marketers acknowledged the Central Six. Different groups' and cultures' consumption propensities would have to be represented as partly overlapping clusters in a six-dimensional space of continuously variable traits. Instead of making superficial generalizations, marketers would have to measure personality profiles across groups by using empirically reliable, valid instruments applied to representative population samples. This would admittedly be more expensive than gathering basic demographic data (age, sex, race, class)—but it would be much cheaper in the long run than misconceived product design and misguided marketing strategy.

Second, marketing would be a much simpler, more stable, more progressive field of human knowledge if it were based on simple, stable, progressive models of humans derived from up-to-date psychology research. But that would be a bad thing for business-school academics seeking tenure through highly citable innovations in consumer research, and for marketing consultants seeking clients for faddish new consumer personality tests and market segmentation techniques. Tying them down too closely to the massive empirical research and solid theory in intelligence research and personality research would leave little wiggle room for radical new books, videos, and training seminars.

As the science historian Thomas Kuhn pointed out, once a science finds a winning formula—a way of making predictable, cumulative progress—it achieves the status of a "paradigm." When paradigms are chugging along happily, and normal science is being done, radical "paradigm shifts" become less likely to succeed. This is as it should be: the more we know about some domain, the less likely it is that a random new idea about the domain will be correct. The more complex the organism, the less likely it is that a random mutation will improve its fitness. The more complex the society, the less likely it is that a major political revolution will improve everyone's welfare.

However, consumer researchers often profit from touting their latest hypotheses as radical paradigm shifts that can offer their consulting clients big advantages over rivals. In order to keep the merry-go-round of continual paradigm shifts dizzying the clients and spinning the money, any upstart theory that looks likely to yield a genuine, stable, unifying paradigm must be killed in its cradle, for the collective good of the consulting profession. This is the economic rationale for why marketing abhors well-established psychological constructs. More scientifically minded market researchers would no doubt welcome the opportunity to understand consumer preferences more deeply using the Central Six, perhaps combined with demographic information. However, because the concepts of general intelligence and the Big Five personality traits are already in the public domain, they can't easily be turned into a patented new business model or other form of intellectual property, so they can't lead to a start-up company that grows faddish and gets bought out.

There is also an intellectual reason for marketers to overlook the Central Six: well-established scientific theories get boring after a while. Indeed, this is a danger of attending too many meetings of the International Society for Intelligence Research: one hears talk after talk about how good old-fashioned measures of good old-fashioned general intelligence predict yet another aspect of human behavior better than any other construct. The same holds true at personality psychology conferences: most talks now identify how the Big Five, yet again, capture most of the human variation in behavior—including the variation that some exciting new measure claims to tap for the first time. Again and again, the Central Six show their reliability and validity in individual differences research—a situation that leaves serious psychologists a little bored, but mostly happy, because we know that we really are making cumulative scientific progress. On the other hand, the stable, ubiquitous power of the Central Six would drive most marketers nuts, because they wouldn't see the individual glory or corporate competitive advantage in using the same methods to describe consumer varia-tion that everybody else uses, or even the same ones they themselves used last year. They want something new and secret: the radical new way to chop the population into chunks that can be optimally targeted by new product lines and advertising campaigns. The Central Six offer no hope of that, because: they are each continuous normal distribu-tions; they have been well understood by psychologists for twenty years; they can be measured with great reliability and validity by exist-ing questionnaires; and they are common knowledge. They offer no excitement, only accuracy; no trendiness, only solidity.

In particular, marketers like to measure individual differences in very domain-specific ways that require arcane knowledge. They mea-sure "individualism versus collectivism" and "abstract versus associa-tive thinking style" to characterize cross-national differences. They measure "masculinity versus femininity" and "gender-role conformity" to characterize sex differences. They measure "strength of reference group influence" to characterize age differences. They measure internal versus external "locus of control" to characterize religious differences. These domain-specific dimensions make life deliciously complex. How dull it would be if we realized that "collectivism" and "strength of reference group" are basically conscientiousness, "abstract thinking style" is general intelligence, "internal locus of control" is quite similar to the surgency facet of extraversion, and "masculinity" often means little more than low agreeableness, low conscientiousness, and high stability.

You might think that individual marketers would want to build solid, accurate models of consumer preferences so that their firms could maximize sales and profits, but you'd be wrong. The social, sexual, and career incentives for individual marketers to be exciting, trendy, and cool are often poorly aligned with the financial interests of a firm's shareholders. In fact, a leading economic theory of advertis-ing suggests that the content of marketing and advertising is largely irrelevant. Rather, the costs that a corporation incurs through market-ing are largely ways for the corporation to signal its financial strength to potential employees, investors, and rival corporations, rather than ways actually to attract customers. This theory of advertising as conspicuous corporate waste follows the same costly signaling logic we have encountered so often before. Insofar as it's true, the real function of marketers is not to understand and influence customers, but to earn high salaries, make a lot of noise, and stay away from the assembly line.

# 11

## General Intelligence

HUMAN INTELLIGENCE HAS two aspects that make it a bit confusing at first. There is a universal aspect: intelligence as a set of psychological adaptations common to all normal humans, including capacities for learning language, using tools, and understanding other people's beliefs and desires. Then there is an individual-differences aspect: intelligence as a set of correlated differences in the speed and efficiency of these natural human capacities, and in our abilities to master evolutionarily novel, counterintuitive concepts and skills such as proving geometric theorems and sustaining lifelong monogamous marriages. Intelligence's universality means that all normal adult humans have some impressive mental capacities for surviving, socializing, mating, and parenting. Intelligence's variability means that some people are much better at these tasks than others.

General intelligence (a.k.a. IQ, general cognitive ability, the *g* factor) is a way of quantifying intelligence's variability among people. It is the best-established, most predictive, most heritable mental trait ever discovered in psychology. Whether measured with formal IQ tests or assessed through informal conversations and observations, intelligence predicts objective performance and learning ability across all important life domains that show reliable individual differences.

The irony about general intelligence is that ordinary folks of average intelligence recognize its variance across people, its generality across domains, and its importance in life. Yet educated elites meanwhile often remain implacably opposed to the very concept of general intelligence, and deny its variance, generality, and importance. Professors and students at elite universities are especially prone to this pseudohumility. They socialize only with other people of extraordinarily high intelligence, so the width of the whole bell curve lies outside their frame of reference. I have met theoretical physicists who claimed that any human could understand superstring theory and quantum mechanics if only he or she was given the right educational opportunities. Of course, such scientists talk only with other physicists with IQs above 140, and seem to forget that their janitors, barbers, and car mechanics are in fact real humans, too, so they can rest comfortably in the envy-deflecting delusion that there are no significant differences in general intelligence.

Even within my own field, evolutionary psychologists tend to misunderstand general intelligence as a psychological adaptation in its own right, often misconstruing it as a specific mental organ, module, brain area, or faculty. However, it is not viewed that way by most intelligence researchers who, instead, regard general intelligence as an individual-differences construct—like the constructs "health," "beauty," or "status." Health is not a bodily organ; it is an abstract construct or "latent variable" that emerges when one statistically analyzes the functional efficiencies of many different organs. Because good genes, diet, and exercise tend to produce good hearts, lungs, and antibodies, the vital efficiencies of circulatory, pulmonary, and immune systems tend to positively correlate, yielding a general "health" factor. Likewise, beauty is not a single sexual ornament like a peacock's tail; it is a latent variable that emerges when one analyzes the attractiveness of many different sexual ornaments throughout the face and body (such as eyes, lips, skin, hair, chest, buttocks, and legs, plus general skin quality, hair condition, muscle tone, and optimal amount and distribution of fat). Similarly, general intelligence is not a mental organ, but a latent variable that emerges when one analyzes the functional efficiencies of many different mental organs (such as memory, language ability, social perceptiveness, speed at learning practical skills, and musical aptitude).

General intelligence seems to be a fairly good general index of genetic quality (not having too many harmful mutations) and phenotypic condition (having a healthy, efficient brain). This is why, across any broad sample of ordinary people, general intelligence correlates positively with:

- overall brain size (as measured in living people by structural MRI)
- sizes of specific cortical areas (such as lateral and medial prefrontal cortex and posterior parietal cortex)
- concentrations in the brain of particular neurochemicals (such as N-acetyl-aspartate)
- age at which the cortex is thickest in childhood
- speed of performing basic sensory-motor tasks (such as pushing buttons as soon as they light up)
- speed with which nerve fibers carry impulses through the arms and legs
- height
- physical symmetry of the face and body
- physical health
- longevity
- semen quality in males (sperm count, concentration, and motility)
- mental health (lower rates of schizophrenia, post-traumatic stress, and other psychopathologies)
- romantic attractiveness (at least for long-term relationships)

For example, I was involved in a couple of studies that examined relationships between intelligence and brain size, and intelligence and body symmetry. The psychologist Lars Penke and I reviewed all fifteen studies we could find on the relationship between general intelligence (as measured with reliable, valid IQ tests) and brain volume (as assessed by MRI brain imaging). In a total sample of 935 normal adults, general intelligence correlated +.43 with brain size—much higher than the correlation previously found between intelligence and external head size (about +.2). We also reviewed eight studies on the heritability of brain size among twins and families, which showed an

average heritability of .91 in a total sample of 2,494 normal adults—
a heritability as high as that found for any other human trait (such as
height), and even higher than the .5 to .7 heritability found for intelligence in mature adults. (Heritability measures the proportion of a
trait's variation across individuals that can be explained by genetic
differences between individuals, so it can range from 0 to 1.) Recent
twin research has also found that there is a positive genetic correlation
between intelligence and brain size, meaning that many genes have
similar positive or negative effects on both intelligence and brain size.
In other words, bigger brains are associated with higher intelligence
not just because the same environmental factors (nutrition, education)
help or harm both, but because the same genes help or harm both. So,
general intelligence and brain size are highly heritable, and they are
moderately correlated at both the trait level and the genetic level. This
is just one piece of evidence suggesting that intelligence is a genuine
individual-differences trait with a deep biological basis.

In another study, psychologists Mark Prokosch, Ron Yeo, and I
studied the relationship between intelligence and body symmetry. We
recruited seventy-eight male college students, and used digital calipers
to measure their left-right symmetry at ten points on their bodies (such
as ankle width, elbow width, ear width, ear length, finger lengths). Body
symmetry is often measured in biology as an index of physical health,
condition, genetic quality, and/or fitness. We also gave the students five
mental tests each (an excellent intelligence test called Raven's matri-
ces, two decent intelligence tests based on vocabulary knowledge, and
two tests of number memory that are reliable, but that are not good
measures of intelligence). We found that, across individuals, scores on
the best intelligence test (Raven's) correlated about +.39 with overall
body symmetry. We also found that the better a mental test was at
measuring intelligence (the higher its "g-loading," in technical terms),
the more highly correlated its scores were with body symmetry across
individuals. Other work by Ron Yeo and colleagues shows that higher
body symmetry is also associated with lower risks of neurodevelopmen-
tal disorders such as mental retardation and lower risks of mental ill-

nesses such as schizophrenia. This suggests that the closer a mental test comes to measuring general intelligence, the closer it comes to measuring general health, fitness, and genetic quality. Again, intelligence appears to be a genuine biological trait with a deep connection to organic processes of bodily growth and brain efficiency.

In the 1970s, critics of intelligence research such as Leon Kamin and Stephen Jay Gould wrote many diatribes insisting that general intelligence had none of these correlations with other biological traits such as height, physical health, mental health, brain size, or nerve conduction speed. Mountains of research since then have shown that they were wrong, and today general intelligence dwells comfortably at the center of a whole web of empirical associations stretching from genetics through neuroscience to creativity research. Still, the anti-intelligence dogma continues unabated, and a conspicuous contempt for IQ remains, among the liberal elite, a fashionable indicator of one's agreeableness and openness.

Yet this overt contempt for the concept of intelligence has never undermined our universal worship of the intelligence-based meritocracy that drives capitalist educational and occupational aspirations. All parents glow with pride when their children score well on standardized tests, get into elite universities that require high test scores, and pursue careers that require elite university degrees. The anti-intelligence dogma has not deterred liberal elites from sulking and ranting about the embarrassing stupidity of certain politicians, the inhumanity of inflicting capital punishment on murderers with subnormal IQs, or the IQ-harming effects of lead paint or prenatal alcoholism. Whenever policy issues are important enough, we turn to the concept of general intelligence as a crucial explanatory variable or measure of cognitive health, despite our Gould-tutored discomfort with the idea.

## Educational Credentialism

You've probably heard that IQ tests are now widely considered outdated, biased, and useless, and that there's more to cognitive ability than general intelligence—there are also traits like social intelligence, practical intelligence, emotional intelligence, creativity, and wisdom. Strikingly, these claims originate mostly from psychology professors at Harvard and Yale. Harvard is home to Howard Gardner, advocate of eight "multiple intelligences" (linguistic, logical-mathematical, spatial, musical, bodily kinesthetic, interpersonal, intrapersonal, and naturalist). Yale is home to Peter Salovey, advocate of emotional intelligence, and was, until recently, home to Robert Sternberg, advocate of three intelligences (academic, social, and practical). (To be fair, I think the notions of interpersonal, social, and emotional intelligence do have some merit, but they seem more like socially desired combinations of general intelligence, agreeableness, conscientiousness, and/or extraversion than distinctive dimensions that extend beyond the Central Six.)

Is it an accident that researchers at the most expensive, elite, IQ-screening universities tend to be most skeptical of IQ tests? I think not. Universities offer a costly, slow, unreliable intelligence-indicating product that competes directly with cheap, fast, more-reliable IQ tests. They are now in the business of educational credentialism. Harvard and Yale sell nicely printed sheets of paper called degrees that cost about $160,000 ($40,000 for tuition, room, board, and books per year for four years). To obtain the degree, one must demonstrate a decent level of conscientiousness, emotional stability, and openness in one's coursework, but above all, one must have the intelligence to get admitted, based on SAT scores and high school grades. Thus, the Harvard degree is basically an IQ guarantee.

Elite universities do not want to be undercut by competitors. They do not want their expensive IQ-warranties to suffer competition from cheap, fast IQ tests, which would commodify the intelligence-display market and drive down costs. Therefore, elite universities have a hypocritical, love-hate relationship with intelligence tests. They use the IQ-type tests (such as the SAT) to select students, to ensure that their IQ-warranties have validity and credibility. Yet they seem to agree with the claim by Educational Testing Service that the SAT is not an IQ test, and they vehemently deny that their degrees could be replaced by IQ tests in the competition for social status, sexual attractiveness, and employment. Alumni of such schools also work very hard to maintain the social norm that, in casual conversation, it is acceptable to mention where one went to college, but not to mention one's SAT or IQ scores. If I say on a second date that "the sugar maples in Harvard Yard were so beautiful every fall term," I am basically saying "my SAT scores were sufficiently high (roughly 720 out of 800) that I could get admitted, so my IQ is above 135, and I had sufficient conscientiousness, emotional stability, and intellectual openness to pass my classes. Plus, I can recognize a tree." The information content is the same, but while the former sounds poetic, the latter sounds boorish.

There are vested interests at work here, including not just the universities but the testing services. The most important U.S. intelligence-testing institution is the Educational Testing Service (ETS), which administers the SAT, LSAT, MCAT, and GRE tests. ETS is a private organization with about 2,500 employees, including 250 Ph.D.s. It apparently functions as an unregulated monopoly, accountable only to its board of trustees. Although nominally dedicated to the highest standards of test validity, ETS is also under intense legal pressure to create tests that "are free of racial, ethnic, gender, socioeconomic, and other forms of bias." This means, in practice, that ETS must attempt the impossible. It must develop tests that accurately predict university performance by assessing general intelligence, since general intelligence remains by far the best predictor of academic achievement. Yet since intelligence testing remains such a politically incendiary topic in the United States, it is crucial for ETS to take the position that its "aptitude" and "achievement" tests are not tests of general intelligence. Further, its tests must avoid charges of bias by yielding precisely equal distributions of scores across different ethnic groups, sexes, and classes—even when those groups do have somewhat different distributions of general intelligence. So, the more accurate the tests are as indexes of general intelligence, the more biased they look across groups, and the more flack ETS gets from political activists. On the other hand, the more equal the test outcomes are across groups, the less accurate the tests are as indexes of general intelligence, the less well they predict university performance, and the more flack ETS gets from universities trying to select the best students. ETS may be doing the best it can, given the hypocrisies, taboos, and legal constraints of the American cognitive meritocracy. However, it may be useful for outsiders to understand its role in higher education not just as a gatekeeper but as a flack absorber. ETS throws itself on the hand grenade of the IQ test controversy to protect its platoon mates (elite universities) from the shrapnel.

If a university degree basically functions as an IQ guarantee, then a degree's social status and economic value should be more strongly predicted by the average SAT scores of graduating students, rather than the average knowledge learned by those students. An IQ-guarantee degree is what economists call a positional good—a way of showing one's personal superiority over competitors. Positional goods often lead to runaway status competition. Once an Ivy League undergraduate degree becomes popular and therefore less useful as a badge of distinction, competitors feel obligated to raise the bar and go for an Ivy League M.B.A., M.D., or Ph.D. Once an ordinary M.B.A. becomes popular and therefore less distinctive, competitors may pursue the more demanding Trium Global Executive M.B.A. for $87,000, in which an elite group of forty senior managers take six trips around the world to study at the London School of Economics, the New York University Stern School of Business, and the HEC School of Management in Paris, plus rotating locations in the Far East and emerging markets.

British universities do not rely so heavily on standardized testing. Instead, their admissions system depends mostly on eighteen-year-old students' standardized test scores in demanding "A-level" courses, which are roughly equivalent to American college sophomore courses. Oxford and Cambridge also challenge applicants' brains and nerves with tough interview questions, such as:

- "What percentage of the world's water is contained in a cow?"
- "Are you your body?"

- "Was Russia just too damned big for democracy?"
- "Why don't we have just one ear in the middle of our face?"
- "What about fatalism?"

Although such questions may not measure intelligence with the same reliability and validity as the SAT, they give interviewers some sense of the applicant's verbal fluency, creativity, and background knowledge. The result is that in Britain, too, an elite university degree functions as an intelligence guarantee.

Credentialism explains the three hundred "diploma mills" (unaccredited online universities) now operating in the United States, such as Rochville University and Belford University. These award B.A.s, M.B.A.s, or Ph.D.s within seven days on the basis of "life and work experience" ("Get a degree for what you already know!"), with no admissions requirements, attendance, classes, essays, or tests. For example, Belford University's "complete doctorate degree package costs only $549 with free shipping," and includes one degree, two transcripts, one award of excellence, one certificate of distinction, and four education verification letters for employers. Available majors include aerospace engineering, clinical psychology, endodontics, and, of course, marketing. Degree mills typically claim accreditation from fronts such as the International Accreditation Agency for Online Universities, which is not one of the nineteen accrediting organizations recognized by the U.S. Department of Education. From 1980 through the early 1990s, the FBI Operation Diploma Scam task force closed down many of these diploma mills. However, there is now virtually no policing of them, and the Web makes it easy to take money from the more gullible careerists of the developing world, to whom $549 sounds like a credibly large amount of money to pay for a Ph.D.

Credentialism also explains the popularity of fake degrees sold through online lost-diploma replacement services such as Bogusphd .com and Noveltydegree.com. For about $50 (including free two-day shipping), you can request any degree from any university, bearing your name, printed on high-quality sixty-pound parchment, with an embossed gold foil seal, suitable for framing. Degree layouts are fairly standard across universities, so they are much easier to counterfeit than a $20 bill. There is the university name at the top, some wording, the graduate's name and degree awarded, some more wording, and then some illegible signatures at the bottom.

An alternative to the credentialist view of higher education is the "human capital" view promoted by most professors, including economists such as Gary Becker. Their starting point is Bertrand Russell's insight that "the average man's opinions are much less foolish than they would be if he thought for himself." The human capital view is that the cultural transmission of knowledge makes us smarter and wiser, and that it is the responsibility of education to download such knowledge into our heads, so that we become more valuable to employers, and earn higher salaries. Education, in other words, is economic self-investment—we forgo some lost wage years in return for higher wages later. And it is a fairly safe investment, barring brain injury or early dementia: Ben Franklin observed, "If a man empties his purse into his skull, no one can take it from him."

The human-capital view argues that education actually confers "added value" on students, making them better workers and citizens who are more useful to society by transforming latent talents into manifest skills and knowledge. A problem with this view is that there are much more efficient ways to learn career-relevant skills and facts: through reading books, watching documentaries, talking with experts, and finding mentors. In *Good Will Hunting*, the title character, a self-educated genius, mocks the Harvard students: "You wasted $150,000 on an education you coulda got for a buck fifty in late charges at the public library." Charles William Eliot, Harvard's president from 1869 to 1909, admitted, "One could get a first-class education from a shelf of books five feet long"—as long as it was the right five feet, such as the fifty volumes of *Harvard Classics* that he edited. The massive rise in homeschooling shows that many parents have come to realize that learning, especially below the college level, need not depend on credentialed schools.

Live college lectures cost about $100 per hour to attend, and are often given by underqualified graduate students or adjunct faculty. Excellent recorded lectures by nationally respected professors on DVD cost about $6 per hour from the Teaching Company (for example, $70 for a twelve-hour course on "Existentialism and the Meaning of Life"). Or, for $550 per year, one could get 450 TV channels from Comcast, even in Albuquerque, including twenty-two nonfiction documentary channels such as the Discovery Channel, Discovery Health, the National Geographic Channel, the Learning Channel, the Travel Channel, the History Channel, the Science Channel, the Military Channel, the Biography Channel, History International, and BBC America. Even better, UK residents can pay their $200 per year TV license fee and see excellent documentaries every evening on BBC1, BBC2, and Channel 4.

Companies understand perfectly well that higher education is not the most efficient way to prepare employees, which is why they spend more than $10 billion per year on corporate training in the United States (versus higher education spending of $200 billion per year). For example, three-quarters of the Fortune 500 companies have bought corporate training from the FranklinCovey group (2007 sales: about $280 million), founded by Stephen R. Covey after his 1989 bestseller *The Seven Habits of Highly Effective People*. By contrast, Harvard's net undergraduate tuition intake (nominal costs minus financial aid) was about $60 million in 2004.

The credentialist view suggests that higher education may offer not so much added value in the economic sense as a reliable method of advertising one's talents—especially intelligence, conscientiousness, and openness. Other credible views of education are also skeptical about its alleged added value. These include the "warehousing" view which holds that mass public education is just cheap child care for working parents, and the "conformism" view, which contends that school socializes children to be reliable, politically pacified wage slaves. Each critique has its merits, and explains why so many students prefer easy courses that one can pass by displaying the correct ideological attitudes (as in some humanities and social sciences) to harder courses that require the acquisition of real skills, knowledge, and insight (as in foreign languages, studio art and music, and the physical, biological, and behavioral sciences). Most students want to maximize their grade point average (a credentialist goal), not the amount of challenging, counterintuitive, life-relevant material they learn (a value-added goal).

This apparently lazy strategy makes sense for most students. Indeed, lingering snobbery often assigns a higher economic value to degrees that are apparently less career relevant. Stanford Law School may prefer an Ivy League history graduate to a pre-law government major from a state college; the BBC often opts for Oxford English degrees over Nottingham degrees in media studies. The highly selective credential with little relevant content often trumps the less-selective credential with very relevant content. Nor are such preferences irrational. General intelligence is such a powerful predictor of job performance that a content-free IQ guarantee can be much more valuable to an employer or graduate school than a set of rote-learned content with no IQ guarantee. This clarifies many otherwise puzzling aspects of higher education, such as the common early-twentieth-century view that "a gentleman need not know Latin, but he should at least have forgotten it." At least my Latin teachers at Walnut Hills High School (Cincinnati, Ohio) were open about why we had to learn to read Virgil: familiarity with Latin roots, prefixes, and suffixes would boost our SAT vocabulary test scores. You know some costly signaling is going on when thousands of teenagers spend three years each learning a long-dead language just so they'll score better on an IQ test that pretends it's not an IQ test, so they can spend four more years and a hundred thousand dollars to get a college degree that pretends it's not an IQ guarantee.

Imagine if we tried to display a physical trait such as aerobic endurance in such a costly, indirect fashion. We could just run naked and barefoot along a five-mile dirt track while others timed us with an Accusplit Survivor II stopwatch ($8.93 retail). But that would be so gauche, so crude, so infra dig. Much better for each person to spend twenty years building a three-hundred-foot-high ziggurat of imported marble, to show that they can run up and down it forty times within an hour—preferably while wearing embroidered silk robes and carrying a solid-gold torch, while a 250-piece marching band plays. This would preserve the rich cultural tradition of Ziggurat Ascension, with its medieval vestments, nostalgic anthems, and bittersweet Sisyphean symbolism. Plus, it would be good for the economy. Parents would have to take out second mortgages to cover their kids' ascension rites. The marble importers, vestment embroiderers, and band musicians would ferociously denounce any reductionistic attempts to measure aerobic capacity with mere dirt tracks and stopwatches. While they might acknowledge that the ziggurat system had some inefficiencies, they would argue that these could be reduced by progress in ergonomically optimized ziggurat stairs, lighter platinum torches, and trombone-playing robots. Contemporary higher education is our ziggurat ascension: an absurdly expensive, time-consuming way to guarantee intellectual and personality traits that could be measured far more cheaply, easily, and reliably by other means. Thorstein Veblen explained most of this perfectly clearly in his 1914 book *The Higher Learning in America*, but, as usual, his insights were nervously appreciated and then promptly forgotten.

## Other Intelligence Indicators

Educational credentials are by no means the only products we buy to display our intelligence. Intellectual narcissism, refracted through consumerism, yields a rainbow of applicable goods and services. We stay informed (capable of impressive dinner-party chatter) by buying news magazines and nonfiction books. We pay the cable TV bill to watch the Discovery Channel. We may buy a Grundig Satellit 800 Millennium shortwave radio for $500 ("Stay alert! Be informed!"), for unfiltered access to foreign news. We take adult-education classes to learn intelligence-demanding leisure activities such as piano playing, oil painting, furniture making, wine appreciation, and conversational Cantonese. We may do some vanity publishing through Xlibris (now a subsidiary of Author Solutions). Or we may be a discriminating collector of high-culture trinkets, like the bidder who bought James Joyce's draft of the chapter "Eumaeus" from *Ulysses* in July 2001 for £861,250 from Sotheby's.

The goods and services marketed to boost apparent intelligence span the whole of human life. Parents encourage children to boost their conspicuous cognitive skills through IQ-building toys, private schools, recreational reading, music lessons, and cultural field trips. For mature adults, displaying a credible level of intelligence is probably a major function of keeping up with news and current affairs, and of reading discussable novels and quotable nonfiction. Older adults may pursue "intelligent retirement" at the Academy Village in Tucson, Arizona. ("For the inquisitive mind, there is no retirement, only more time to learn, explore, share, and grow.") Also, as older travelers grow more numerous, richer, and more sophisticated, new market niches open up for exotic regimens of intellectual self-improvement and self-display. There are expeditions with Smithsonian Study Tours, or Far Horizons Archaeological and Cultural Trips, and small-group cultural tours with Martin Randall Travel, such as "Gastronomic Catalonia," "Connoisseur's Vienna," or "Mediaeval Normandy."

Many intelligence-indicating products require mastery of arcane knowledge and technical specifications. We may get into home astronomy to show off our technical savvy and sense of wonder at the myriad galaxies, preferably as viewed through the best-value 140 APO Refractor with six-inch-diameter tube and FeatherTouch focuser ($4,750 from the Telescope Engineering Company), or, even better, the C-400 Classical Cassegrain with 400 mm aperture ($50,000 from Takahashi)—which can take a good photo of a quasar 10 billion light-years away. Such telescopes become props for showing off one's astronomical knowledge. Similarly, the GammaMaster watch with built-in Geiger counter may not seem practically useful to consumers other than Homer Simpson or James Bond, but it allows one to talk about the reasons why being exposed to more than one thousand micro-sieverts per hour of gamma radiation is likely to cause mutations that undermine the quality of one's natural fitness indicators.

Even products that used to be simple have now acquired such complex features that they have become reliable intelligence indicators. This "feature creep" (the ever-increasing number and complexity of controls and features) is driven partly by the need to make each new product model different from last year's, but also partly by the consumer's unconscious desire for a product that is right at the limit of his cognitive ability, and one that therefore functions as a credible cognitive display. Consider sewing machines. The iconic Singer Class 201 machines (circa 1939–1950s) were hefty black metal appliances with a couple of dozen moving parts and a variable-speed pedal switch. By contrast, the recent Janome Memory Craft 11000 sewing machine ($7,500 MSRP) is basically a computerized desktop sewing robot controlled by a 7.5-inch VGA touchscreen. It includes two USB ports for downloading .JEF format embroidery designs from a home PC, which it can then copy, paste, flip, rescale, and rotate using up to one gigabyte of onboard memory, before it sews them at up to eight hundred stitches per minute across surfaces up to 8" by 11". It can also produce 307 preprogrammed stitch patterns, three lettering fonts, and thirteen buttonhole patterns. Although easy and intuitive to use for basic sewing tasks, it would take an extraordinary mind to fully exploit its capabilities.

More adventurous techno-enthusiasts may get a private pilot's license, which requires intellectually demanding ground training (flight planning, navigation, aerodynamics, air traffic control rules, weather theory), flight training, and about $100 per hour of airplane rental time. If a private pilot has survived a few thousand flying hours, you can be pretty sure he or she is highly intelligent, conscientious, and emotionally stable. Evolution in action: the less intelligent and diligent pilots fly into box canyons, crash, and die.

Perhaps the purest intelligence-indicating products are strategy games, broadly construed. We buy cards, Go boards, chess sets, and the *New York Times* for the Will Shortz crossword puzzles. When Gary Kasparov was narrowly defeated by IBM's Deep Blue computer in 1997, it was nonetheless impressive that his expertise and intuition could match the computer's ability to analyze 200 million board positions per second. We may try to emulate his mastery by training with Deep Shredder 9 from ChessBase ($50 on CD-ROM), which gives access to 2.7 million games played by grand masters. The current fad for public-transit commuters to do sudoku puzzles is perhaps the purest and most public display of analytical reasoning ability ever devised. Unlike crosswords, sudoku requires no verbal background knowledge; unlike chess, it can be played alone; unlike gossiping publicly on one's cell phone, it requires intelligence.

Active online equity trading (day-trading) is an even more challenging, higher-stakes intelligence-indicating game. Every rational investor knows that the stock market is virtually random in its movements, and that the most reliable long-term bet is an index-tracking fund with annual expenses under 0.5 percent. It's hard to beat the market, because large investment banks can use insider information, employ highly paid experts, and use automated trading software that can analyze hundreds of variables. Because each online trade by a private investor costs about $10, the expected return from active trading is negative. Nevertheless, many investors try to outsmart the market through their own research. Why? Because it's a thrilling intellectual game. If you consistently make money from risky, highly leveraged positions, on the basis of your own research, then some other suckers have lost money because their research, analysis, judgment, and self-control were inferior. It's like successfully bluffing in poker. Your brains and your guts beat theirs. Rather than viewing active equity trading as nothing more than a symptom of runaway greed and macho aggressiveness (low agreeableness), we should also view it as a costly, risky intelligence indicator.

Marketers have developed some intuitions about how to inform consumers that a product has intelligence-signaling potential, even though the connotations of "intelligence" are so problematic that the word is rarely used in ad copy. The preferred euphemism is the affix "smart," as in:

- Smartfood: Frito-Lay's white-cheddar-cheese-flavored popcorn
- Smartwater: Glacéau's electrolyte-enhanced bottled water
- Smart Start: Kellogg's breakfast cereals
- Smart Car: the small European car with the three-cylinder 1.0 liter engine
- *SmartMoney:* the personal-finance magazine
- Smart Bar: the cutting-edge Chicago bar and dance club
- Smart Parts: the online retailer of paintball products, such as Exoskin knee pads and the Ion XE gun, the "first to deliver true electropneumatic performance in the casual paintball player's price range"

More subtly, the iPod and the BMW 500i both contain the letter *i* to suggest the intelligence of their users. Marketers more often than not confuse intelligence with wealth, status, taste, class, or education, and don't understand the distinctive product attributes that bright consumers actually seek to display their brightness. Marketing types don't understand that tech-savvy males unconsciously respond to product technical specifications not as useful features they'd enjoy, but as impressive properties that they'll be able to talk about in impressively articulate, IQ-displaying ways. A bag of Smartfood popcorn, once opened, cannot be put aside by any normal human until it is fully consumed, so it provides an opportunity to display one's intelligence in avoiding salt-induced hypertension by not opening the bag in the first place. The iPod provides an opportunity to explain the meaning of "160 GB" and "MPEG-4" (in case you do, in fact, know what these mean). The BMW 550i provides an opportunity to display one's mastery of the cortex-sapping iDrive system and the built-in iPod and iPhone docks, and to explain the meaning of "Double-VANOS steplessly variable valve timing," "Servotronic variable-ratio power steering," and "dynamic auto-leveling Xenon Adaptive Headlights." For each such product, the consumer—usually male, in such cases—tracks the scent of cognitive complexity. He thinks *Those features sound awesome,* which roughly translates, "Those features can be talked about in ways that will display my general intelligence to potential mates and friends, who will bow down before my godlike techno-powers, which rival those of Iron Man himself."

A final way to boost one's status through intelligence-indicating products is to rent the intelligence of others. Typically this has involved hiring craftsmen of very rare, very high intelligence to create custom works of exquisite complexity and novelty. For example, medieval Muslim rulers, priests, and traders competed for status partly by commissioning architecture with very complex abstract ornamentation, especially *girih* (geometric star-and-polygon) patterns. There was apparently a runaway competition in the complexity of *girih*-design, such that by the fifteenth century AD, Muslim master builders were constructing almost perfect quasi-crystalline Penrose tiling patterns, as in the Darb-i Imam shrine in Isfahan, Iran. These patterns were so complex and required such intelligence to design that they were not rediscovered in European mathematics until the 1970s by Roger Penrose. Intricate, inventive, custom products bring status to the purchaser, because the bell curve of intelligence guarantees that in any given culture, the genius required to design such products is rare, is in great demand, and hence is very expensive. The wealthy have always recognized the value of commissioning works from the greatest geniuses they can patronize.

## Intelligence-Boosting Products

Another class of products promises to enhance our apparent intelligence in various ways. Intelligence tends to peak in young adulthood, at the height of social ambition and mating effort, as manifest in the young-adult outpouring of creativity in music, art, and humor, which all require a quick-witted facility with novel problem solving. Thus, products to boost intelligence tend to be especially sought by young adults who aspire to higher intellectual status and creativity.

These intelligence boosters include many of the thought aids that help adapt our prehistoric crania to the challenges of modern living. They include decision aids (calculators, spreadsheets, expected utility theory), time-allocation aids (watches, calendars, diaries), communication aids (maps, books, phones, e-mail, PowerPoint), and social reciprocity aids (money, invoices, checks, debit cards). Such products are like cognitive crutches.

Other products aim to boost human mental abilities beyond the normal. These include the Mozart Effect CD collection that allegedly "can help you achieve the kind of measurable IQ boost documented in the famed University of California at Irvine study." The Baby Einstein and Brainy Baby brands make similar claims that infants can benefit cognitively from lines of videos, music, books, toys, and games. Less explicitly, for decades toy brands have implied that products for constructive play (LEGO, K'Nex, Erector, Brio), imaginative social play (Playmobil, Imaginarium, Cranium), and game play (ThinkFun, University Games, Imaginext, Harvard, Edu Science) will promote children's cognitive development. The Nintendo DS game Brain Age claims to boost cortical function through memorization exercises, puzzles, and math. Websites such as Happy Neuron and MyBrainTrainer offer similar services. All such products are marketed, more or less, as mental superchargers.

Intelligence boosters also include some drugs. Exam-cramming students of the twentieth century favored the older generation of psychoactive drugs (caffeine, nicotine, cocaine). Nowadays, online-game players seeking faster sensory-motor reactions and sustained tactical attention prefer the newer generation of "energy drinks" and "smart drugs" (containing taurine, ginseng, ginkgo biloba, or creatine). White-collar professionals also use prescription drugs such as Provigil, Ritalin, and Adderall to boost their intelligence and attentiveness.

The new "transhumanist" movement has been especially thoughtful and optimistic about the long-term possibilities of intelligence-enhancing products. The World Transhumanist Association, founded by the philosophers Nick Bostrom and David Pearce, has advocated a basic moral right for people to extend their physical and mental capabilities through any available technologies—life extension, cybernetics, genetic engineering, stem cells, nanotechnology, cognitive enhancement. However, transhumanism has not yet grappled with the behavioral issues that are likely to arise when such technologies hit the consumer retail market. One prediction is obvious: even for commercialized transhumanist products that offer nothing more than a placebo effect, early adopters will acquire and display them mainly as indicators of intelligence and openness. It won't matter whether the first ten-terabyte nanoneural implant actually does make the customer smarter; as long as the implant is expensive, exclusive, well marketed, and clearly branded, it will sell as a costly, conspicuous, limited-reliability signal of high intelligence.

# 12

## Openness

OPENNESS TO EXPERIENCE is a piquant trait, delectable to some but disgusting to many. Its evolutionary origins are obscure, its dangers are often underestimated, and the environmental variables that predict it are strangely mundane. Yet it has prodigious importance in creativity, courtship, and cultural progress.

Wherever you stand on the openness spectrum, those less open than you seem boring, dull, conventional, and conformist, whereas the more open seem eccentric, bizarre, disruptive, threatening, or even psychotic. Given this diversity of openness levels, and the resulting diversity of preferences for different degrees of openness in family, friends, and mates, there is less incentive for people to fake their openness than their intelligence. If you pretend to have higher openness than you really do, you may be transiently attractive to those more open than you, but you'll be less attractive to those less open than you. The net result will be no higher social or sexual popularity. In fact, given the bell curve of openness, the more you deviate from an average level of openness, the fewer people you are likely to attract.

This means that where openness is concerned, consumers can sometimes display cheap, reliable openness badges that are fairly credible. It also means that consumerist capitalism caters very well to the whole range of openness. There are highly open cities (Vancouver, Amsterdam, Bangkok), and less-open suburbs (Langley, Virginia; Wimbledon, London; Sandton, Johannesburg). There are open neighborhoods (wherever gay people and grad students live openly) and more closed-minded neighborhoods (wherever cardiologists live). There are some highly open musical genres (indie, alternative, jazz, world, hip-hop) and more conventionalized genres (pop, country, gospel, classic rock). There are more open genres of fiction (contemporary, science, erotic) and more conservative forms (romance, mystery, military history, fantasy). There are highly open magazines (*Seed, Wired, Prospect, Icon, Harper's, Unzipped*), and more mainstream magazines (*Time, Money, Stuff, Today's Christian Woman*). There are highly open U.S. colleges (Reed, Hampshire, Barnard) and much less open colleges (Princeton, West Point, the Citadel). In every market for every kind of product, the highly open feel at home whenever the unknown surrounds them, whereas the less open want nothing to shake them out of their comfortable stability.

## Why Parasites Reduce Openness

Some exciting new research shows that a surprising environmental factor—the risk of parasitic infection (or "parasite load")—predicts lower openness, lower extraversion, lower individualism, and lower liberalism across individuals and societies. This research, led by biologists Corey Fincher and Randy Thornhill at University of New Mexico, and Mark Schaller and Damian Murray at University of British Columbia, illuminates the nature of openness and the reasons why it varies across individuals.

The scientific team's reasoning is that before modern sanitation and medicine, the most important causes of human disease, death, and infertility were parasites (colloquially, "germs"; technically, "pathogens"). All large-bodied species of animals and plants are in a constant state of biological warfare against dozens of kinds of smaller-bodied parasites. The big problem is that the parasites—viruses, bacteria, protozoa, and worms—are so minuscule and short-lived that they can evolve very quickly—much faster than their larger, longer-lived hosts, such as humans. Some of these alien organisms evolve to form symbiotic relationships with their hosts. These include most of the 100 trillion or so microbes that live in the human gut, which collectively contain one hundred times as many different genes as the human genome does. However, many of these alien organisms remain evolutionary enemies with harmful effects on health and reproduction.

In response, vertebrate animals have evolved a system of biochemical defenses that can learn, during the host's own lifetime, to recognize and attack dangerous local parasites if they enter the host's body. This defense system, which is called the adaptive immune system, is composed mainly of the 2 trillion lymphocytes in our blood and lymphatic system, which collectively weigh as much as the human brain. Each person's lymphocytes learn to fight off the particular varieties of parasites that are common within his own local group, which gives him an immunological memory of the parasites he has already encountered. (Immunization is simply the process of teaching the lymphocytes about a new kind of pathogen by exposing them to safer, deactivated forms of the pathogen.) However, the immune system's learned parasite resistance is highly localized. People from other kin groups, clans, tribes, ethnic groups, or races—even if they live just a few miles away—may host other varieties of parasites that evolved slightly different ways of being transmitted to hosts, infecting them, and making them sick.

Thus, any interaction with outsiders brings a high risk of acquiring a new kind of parasite that may be especially hard for one's locally adapted immune system to fight off. The higher the parasite load—the greater the number, variety, and severity of parasites surrounding one's local group—the higher that risk is and the more cautious people should be about strangers. They should develop a more proactive "psychological immune system" to avoid getting their mouths, noses, genitals, or skin anywhere close to potential sources of infection. They should be much more averse to contact with other groups, including not just their human members, but also their food, clothing, shelters, animals, social customs, hygiene practices, and purification rituals—anything associated with possible parasite transmission. In other words, people in high-parasite regions will benefit from becoming more xenophobic (fearful of out-groups) and ethnocentric (focused on their own in-group). (This effect is so robust that it even occurs in virtual worlds, as in the "Corrupted Blood" plague that ravaged World of Warcraft in September 2005. Some infected "griefers" intentionally spread the plague to highly populated areas of the game-world, killing up to half the characters on some servers; players quickly learned to avoid major cities.)

On the other hand, if the environment is cold, dry, or otherwise hostile to parasite transmission, the parasite load will be lower, the risks of communicable disease will be lower, and the costs of interacting with outsiders will be lower. Since interaction with out-groups can bring fitness benefits—new trading opportunities, resources, knowledge, inventions, mates, friends, allies, and genes—people in low-parasite regions should evolve or learn to take more advantage of these benefits by acting more cosmopolitan. They should evolve or learn to act less xenophobic and ethnocentric as groups. They should also grow more open and extraverted as individuals, because higher openness drives people to seek out new ideas, experiences, places, and cultures, and higher extraversion drives people to seek out new mates, friends, and allies.

In an extraordinary series of recent papers, Fincher, Thornhill, Schaller, and Murray have been confirming all these predictions. Their methods were fairly simple: for each of ninety-eight territories (nations and city-states) selected from around the world, they added up the medically documented prevalences of nine parasites already known to reduce human reproductive fitness (leishmania, schistosoma, trypanosoma, malaria, filaria, leprosy, dengue, typhus, and tuberculosis). Then they looked at whether this measure of parasite load predicts political and social attitudes and Big Five personality traits across the different territories.

For example, in a 2008 paper, Schaller and Murray suggested that openness and extraversion would be lower in territories where people suffer from higher parasite loads. They gathered data on average Big Five personality scores from three previous studies that had each analyzed thirty-three, fifty, or fifty-six of the ninety-eight territories for which the parasite loads were known. Across the seventy-one territories for which they had both parasite-load data and personality data, they found that people from territories with the highest parasite load indeed had substantially lower openness and extraversion scores on average. For the twenty-three territories where Big Five scores could be averaged across all three previous studies (yielding the most accurate estimates), the correlations were quite strong: −.6 between parasite load and openness, and about the same for extraversion. These correlations remained substantial even after controlling for differences across territories in average annual temperature, distance from the equator (absolute latitude), life expectancy, GDP per capita, or political attitudes (individualism versus collectivism). Schaller and Murray suggested that these personality shifts could happen in three complementary ways: evolutionary changes across generations in the genes underlying personality, environmental influences on the expression of personality genes during individual development, and/or cultural changes across historical time in the norms governing social interaction.

In another 2008 paper Fincher, Thornhill, Schaller, and Murray analyzed data on the "individualism versus collectivism" dimension, the political attitude variable that has been most thoroughly studied in crosscultural and political psychology. Collectivists make stronger distinctions between in-group and out-group, are warier of contact with strangers and foreigners, and highly value tradition and conformity. Relatively "collectivist" cultures include China, India, and countries in the Middle East and Africa; relatively "individualist" cultures include the United States and the nations of western Europe, especially Scandinavia. The researchers gathered data on average individualism/collectivism scores from four previous studies that had each analyzed sixty-eight, fifty-eight, fifty-seven, or seventy territories for which parasite loads were known. Across the ninety-eight territories, the various measures of collectivism correlated strongly with current parasite load (correlations ranging from .44 to .59), and even more strongly with historical parasite load from about a century ago (correlations ranging from .63 to .73). (This makes sense: a territory's collectivism level is likely to show some cultural lag when responding to rapid reductions in parasite loads due to modern sanitation and medicine.). Even controlling for the four variables known from previous research to predict collectivism across cultures—life expectancy, population density, GDP per capita, and the Gini index of economic inequality—parasite load still predicted collectivism quite strongly. (In the United States, "collectivism" means being Republican, Fundamentalist, pro-military, and/or anti-immigration; "individualism" means being Democratic, secular, internationalist, and/or antiracism. In the United Kingdom collectivists tend to be Conservative or old Labour, Anglican or Muslim, pro-royalty, and/or anti–European Union; individualists tend to be Liberal Democrat or New Labour, secular, viewers of BBC2 and Channel 4, and pro-EU.)

Does vulnerability to parasites also predict differences in openness, extraversion, and individualism across individuals, even within the same culture? To put a finer point on it: do U.S. Republicans and U.K. Conservatives hold the values that they do, not as a function of rational political convictions, but because they unconsciously assess that they have weaker immune systems? Some recent studies suggest so. The evolutionary researchers Dan Fessler, David Navarette, and Mark Schaller have found that "perceived vulnerability to disease"— an individual's self-rated susceptibility to getting colds, infections, and communicable diseases—does predict that individual's xenophobia. Also, looking at photographs of parasites and disease symptoms has been shown to make people more xenophobic, at least temporarily. A final piece of evidence relies on the fact that women's immune systems grow adaptively weaker during first-trimester pregnancy, so that their bodies don't reject the fetus as an alien parasite. Women in the first trimester also show higher xenophobia, as if they unconsciously realize that their weaker immune systems will have more trouble fighting off new infections from outsiders; this xenophobia becomes weaker as their immune systems become stronger in the second trimester. More generally, people's openness, extraversion, and individualism tend to peak in young adulthood when their immune systems are strongest, and tend to decline throughout middle age as their health declines.

All this research suggests that there are surprisingly strong connections between parasite loads, immune systems, personality traits, and political attitudes—connections that have profound implications for understanding the nature of openness, the reasons why it varies across people and cultures, and the ways that consumers try unconsciously to display their levels of openness. These connections also have surprising implications for economic and political policy. For instance, the best way to promote democracy, science, secularism, peace, ethnic harmony, free trade, and tourism in developing nations may be to reduce their parasite load through better medicine and sanitation. As foreign policy tools, immunizations and mosquito nets may be much stronger than diplomacy, trade agreements, economic sanctions, loans, warfare, or propaganda, because these simpler measures increase openness, extraversion, and individualism more effectively, and reduce xenophobia and ethnocentrism more quickly. Médecins sans Frontières and the Bill and Melinda Gates Foundation may be promoting the cultural liberalization of developing nations more effectively than the UN, WTO, and World Bank combined. Indeed, Randy Thornhill has argued that this liberalization-through-parasite-reduction is exactly what happened in the United States and Europe during the twentieth century. The baby boomers, for example, were not only the first generation in the United States to grow up benefiting from broad-spectrum childhood immunizations, but were also the first generation to show a sudden massive increase in interracial tolerance, internationalism, openness, and individualism, as manifest in the civil rights movement, the peace movement, the New Left, the psychedelic revolution, the sexual revolution, and the "Me" generation. The conservative backlash benefited from the disco-era genital herpes epidemic and especially from the 1980s AIDS crisis, which increased everyone's perceived vulnerability to communicable disease.

## Four Forms of Disgust

As we have seen, low openness has some biological benefits in avoiding new infectious diseases from unfamiliar people, groups, foods, and hygiene practices. These benefits are derived not just from a cool-headed distaste for the new and the different, but by a hot-blooded disgust at the bizarre and the alien. To better understand openness, it may help to consider the nature of disgust.

The evolutionary psychologist Joshua Tybur argues that humans experience three distinct forms of disgust. There is antiparasite disgust, which protects us from contagious disease. There is sexual disgust, which protects us from mating with individuals who are too closely related, or whose genetic quality is too low. Finally, there is moral disgust, which protects us from selfish individuals who would undermine local social norms and social contracts. Each type of disgust is triggered by different stimuli (rancid meat, lustful siblings, or psychopaths) and provokes different behaviors (vomiting, sexual aversion, or vigilante justice).

I think there may also be a fourth type of disgust, a sort of "cultural disgust," that protects us from contact with infectious memes that may prove dangerous given our psychological vulnerabilities. This is the sort of disgust that was aroused in low-openness people by modern art in 1930s Berlin and by tie-dyed T-shirts in 1960s Texas. It is precisely the sort of disgust exploited by rabid conservative talk-show hosts, who are not just frightened by pathogens, incest, and psychopaths, but by certain ideas that they consider dangerously attractive, memorable, transmissible, and psychosis inducing. They portray the carriers of these ideas as crazy liberal zombies, tainted by an ideological epidemic beyond their control, who must be put down before the contagion spreads.

People who (unconsciously) understand that their own immune system is vulnerable—the sick, the elderly, pregnant women in the first trimester, inhabitants of areas with high parasite loads—typically experience stronger antiparasite disgust. Likewise, people who unconsciously understand that their own mental health is vulnerable may experience stronger cultural disgust and lower openness. Whenever you hear a conservative broadcaster such as Bill O'Reilly ranting against some new idea, imagine him as afflicted by the cognitive equivalent of severe combined immunodeficiency syndrome. (This is the weak-immune-system condition that forced David Vetter, the "Bubble Boy," to stay at home and live in an antiseptic plastic tent, lest he be killed by the germs that we routinely shrug off.) O'Reilly might be rightly intuiting that if he seriously considered a wider range of novel ideas, his sanity might collapse. (Or, his ideological defenses may be miscalibrated and he may have become overvigilant against ideological infection, in a way more analogous to autoimmune diseases, such as allergic asthma.)

Conspicuous displays of immune system strength are fairly common across cultures. Especially in areas with high parasite loads, many tribal people open their skin to infection when they are young adults, through scarification or tattooing or forced genital mutilation, to show that their immune systems are strong enough to survive the wounds. (The 5,300-year-old body of "Ötzi the iceman," discovered in the Italian Alps in 1991, had fifty-seven tattoos.) If you are a healthy, energetic young man or woman covered in well-healed self-inflicted scars, despite living in a high-parasite area, you have credibly demonstrated that your health is very strong. Potential mates and friends may not consciously understand the connections between costly signaling theory, microscopic parasites, scarification using unsterilized tools, and individual differences in the number and efficiency of the lymphocytes that constitute the adaptive immune system. However, they can unconsciously assess that you would not be looking healthy or energetic after having endured so many cuts if you were weak and sickly. Biologists such as W. D. Hamilton and Anders Møller have argued that in many other animal species, sexual ornaments have evolved as indicators of parasite resistance.

The phenomenon of "cutting" among American teenagers baffles parents and doctors, but may be an example of this sort of ritual scarification. It rarely has suicidal intent but does often lead to infections, and teens who have few other ways to raise their social status often say it makes them feel cool. As with almost all human behaviors, the environmental risks factors for cutting (poverty, unemployment, uncaring parents, childhood abuse, romantic heartbreak) and the conscious motives for cutting (feeling sad, lonely, numb, lost, alienated, over-whelmed) may have little relation to the evolutionary origins or hidden adaptive functions of the behavior, except insofar as they reveal that the behavior is a high-risk, last-ditch tactic for boosting one's social salience to parents or mates, or one's status among peers. (Only about 10 percent of teens report having cut themselves.) Tattoos, earrings, body piercings, and other self-inflicted breaches of the skin may function in a similar way as a reliable display of immune system strength. Of course, to suggest that cutting and body piercing arise from evolved human instincts to display immune system strength is not to argue that they are either morally good or medically advisable.

In developed countries, we have less to fear from infectious parasites, but much more to fear from infectious memes. So, instead of opening our bodies to ambient germs, we open our minds to ambient culture, to determine if we can stay sane throughout the onslaught. When you see teenagers and young adults posting their interests in music, books, and film on their MySpace websites, consider the costly signaling principles at work. If they have exposed themselves to a lot of death metal, Chuck Palahniuk, and David Lynch, and they are still sane enough to sustain a reasonable conversation through e-mail or instant messaging, they have credibly proven their openness and psychosis-resistance. The main danger is that if they only recently got into such extreme forms of culture, the psychosis could still hit in the next few months, following other stressors. An individual with a long-standing appreciation of Lynch's *Blue Velvet* (1986) is therefore a safer bet than one with a newly acquired enthusiasm for *Inland Empire* (2006).

Certain extreme ideas may present minimal danger to those with strong antipsychosis defenses, who can therefore afford to act highly open. But those same ideas may present genuine dangers to those with weaker defenses, who must minimize their openness. If these outlandish speculations have any merit, then people who are low in openness prefer to associate with one another in part to protect their sanity. They seek out communities, jobs, lifestyles, mates, friends, and products that will not challenge their antipsychosis defenses. They prefer the familiar to the novel, the conventional to the radical, the predictable to the challenging. They prefer goods and services that are heavy on matter and habit, and light on cognition and imagination. They move to comfortably anti-intellectual communities: rural towns or ethnically homogeneous suburbs around provincial cities, such as Indianapolis, Indiana, or Augsburg, Germany; large, progressive, multicultural cities are just too threatening. They shop in national chain stores and eat in chain restaurants, because locally owned businesses are too quirky and unpredictable. They minimize their exposure to "serious culture"—that is, weird, potentially infectious memes—and spend their leisure time soaking up formulaic TV series, romantic comedies and action movies, romance and military-thriller novels, local newspapers, and conventional religious services. In this way, the less-open can thrive for years in meme-excluding bubbles, avoiding as much as possible disturbing thoughts and social encounters. For them, the unexamined life is . . . the easiest way to avoid psychosis.

## Why Don't We All Want Maximum Openness?

Openness is a dangerous trait in several ways. It can lead to social embarrassment when one's behavior is too weird or novel. It can lead to one's brain getting infected by maladaptive memes—false information, dumb ideologies, conspiracy theories. It can lead one to join a cult, enroll in art school, or move to Santa Fe.

For example, while openness is strongly correlated with creativity, it is also correlated with psychosis (loss of contact with reality). To study these correlations, my colleague Ilanit Tal and I asked 225 University of New Mexico students to complete six tests of verbal creativity and eight tests of drawing creativity, along with measures of the Central Six traits. Our six verbal creativity tasks included questions like: "Imagine that all clouds had really long strings hanging from them—strings hundreds of feet long. What would be the implications of that fact for nature and society?" and "If you could experience what it's like to be a different kind of animal for a day, what kind of animal would you want to be, and why?" For the eight drawing creativity tasks, students were asked to create four abstract drawings (for example, "Please draw an abstract symbol, pattern, or composition that represents the taste of pure, rich, dark chocolate") and four representational drawings (for example, "Please draw what an alien civilization might look like on a distant planet"). The students' responses were rated independently for their creativity by four expert raters, who showed very good agreement with one another.

The students also answered seventy-four questions about whether they had experienced various symptoms of schizotypy—a form of mild schizophrenia. People with schizotypy tend to have both "positive" symptoms (delusions, hallucinations, a belief that people can read their mind, odd speech, and odd behavior), and "negative" symptoms (flat emotions, few friends, social anxiety, paranoia). For example, people tend to get a high score on schizotypy if they say yes to many of these positive symptom items:

- I often hear a voice speaking my thoughts aloud.
- Some people think that I am a very bizarre person.
- I believe in telepathy.
- People sometimes stare at me because of my odd appearance.
- Parts of my body sometimes seem unreal or disconnected.

(Most college students answer yes to at least a few such items, so if you did, too, don't panic.)

We found that openness has moderate positive correlations with both general intelligence (.30) and positive schizotypy (.29). Openness predicts both verbal creativity (.34) and drawing creativity (.46). Intelligence also predicts both verbal creativity (.35) and drawing creativity (.29). Schizotypy predicts creativity very weakly, but once you control for openness, it does not predict creativity at all. Thus, creativity is best predicted by positive responses to openness questions, rather than schizotypy questions.

The implication is that there is a link between madness and creativity, as philosophers have speculated for millennia, but the link is mediated by openness. Openness has creativity benefits and extreme openness shades over into psychosis, but the psychosis isn't generating any creative work; it's just a harmful side effect. This is yet more evidence that very high openness is a dangerous game, with potentially high payoffs in creativity, but potentially catastrophic effects on mental health. In a complex, media-rich society, perhaps only people with very good mental health can tolerate a high degree of openness without losing their equilibrium.

It is important to remember that openness is not correlated positively or negatively with emotional stability. High openness is associated with mild psychosis, whereas low stability is associated with neurosis—anxiety, worry, depression. The speculation here is that high openness plus high stability and high intelligence yields high creativity and social attractiveness. Conversely, high openness plus low stability yields a high risk of mental illness.

## How Much Openness Can You Take?

These findings suggest a hypothesis that I admit is highly speculative, but that is nonetheless intriguing: people may use conspicuous displays of openness as a guarantee of their mental health—especially their resistance to developing schizotypy, schizophrenia, or other forms of psychosis. Because schizophrenia tends to develop in early adulthood, it is especially important for adolescents and young adults to display their psychosis resistance. If you're a young woman, you want to avoid falling in love with a guy when he's seventeen, having two children with him, and then seeing him develop debilitating schizophrenia when he's twenty-three. So, adolescents play with fire, exposing themselves to ideas, experiences, and drugs that would induce psychosis if they did not have good psychosis resistance. They test themselves to extremes, and though they sometimes misjudge and go psychotic, mostly they do not. Young adults take great pride in being able to withstand bizarre music, mind-bending books, intense movies, ultraviolent computer games, and powerful hallucinogens. To be cool is to survive such experiences mentally intact. Even older adults often take pleasure in exaggerating the psychosis-inducing severity of quotidian inconveniences—as when a delayed flight or tedious party is described as "a horrific nightmare that almost drove me bonkers."

The entertainment industry obliges by providing a cornucopia of potentially psychosis-inducing products for consumption by the highly open: Films by Quentin Tarantino and David Fincher. Novels by Jeff Noon, Salman Rushdie, and Ursula Le Guin. Music by Beck, Tricky, and the Gorillaz. Courses on Foucault, Derrida, and Baudrillard. Brown cafés in Amsterdam and hookah lounges in Austin, Texas. Holidays in Las Vegas. The Burning Man festival. Art at the Whitney Biennial. Raves and nightclubs on Ibiza. Nonstop eighteen-hour flights from Vancouver to Sydney in a midrow coach-class seat. If you are a young single adult, and expose yourself to enough of these in quick succession, you will gain either an increased level of psychosis, or increased social status and sexual attractiveness among your peers. (Research shows that stressful life events such as these do not by themselves cause full-blown schizophrenia, but they can increase psychotic symptoms among those who already have a genetic risk of schizophrenia, and even among those who do not.)

The highly open expose themselves to new experiences, cultures, people, relationships, norms, ideas, worldviews, art, music, sexual practices, and drugs. They can get infected by nasty, maladaptive memes; they might end up believing in astrology, homeopathy, or Scientology. They might find themselves joining the open-marriage scene, which almost always leads to divorce; or the methamphetamine scene, which leads to psychosis; or both, which leads to spousal homicide. Cultural disgust to bizarre new ideas protects low-openness people not only from psychosis, but from maladaptive memes. They may not adopt useful new ideas very quickly, but neither do they join suicide cults. Their behavior is nicely constrained by time-tested traditions that must, on average, have yielded decent reproductive success.

My own personal openness is especially high, so given that bias, this chapter may have made high openness sound biologically fitter, psychologically healthier, or morally superior. In truth, openness is very different from intelligence. High intelligence is almost always a good thing, because it predicts better performance in almost every domain of life. We can't all achieve high intelligence, however, because most of us carry too many mutations that disrupt brain function. By contrast, high openness is risky, costly, and often maladaptive. Its yields benefits only if the environment is especially safe (low parasite load, low proportion of harmful memes) and if one's defense systems (immune system, mental health) happen to be especially strong. Otherwise, better safe than sorry.

## The Embarrassment and Danger Costs of Openness

Psychosis is not the only danger of openness. High openness, in my experience, can often get one into embarrassing situations that may lower one's social and sexual status, or even threaten one's safety. The highly open may emerge soaked in fake blood from a Gwar concert and be mistaken by police for a homicidal maniac. They may suffer burns and hearing loss from attending an illegal performance by the giant flame-throwing robots of Survival Research Labs under Highway 101 in San Francisco. They may be chased through the streets of Fez by furious temple guards for having kissed an English girlfriend in the Karaouine mosque. They may become evolutionary psychologists, which typically leads to unemployment and ostracism.

Openness to new experiences can even lead to physical danger or addiction. Athletic novelty seekers may run out of ordinary sports to try, and turn to "Xtreme" sports—base jumping, motorcycle ice riding, climbing K2, heli-skiing. (Those low in openness respond with parodies of such endeavors, like "extreme ironing" or "extreme croquet.") Cultural novelty seekers run out of safe countries to visit, so turn to extreme-tourism adventures in Afghanistan or Rwanda. Sexual novelty seekers may tire of safe hetero vanilla sex, and turn to bisexuality, polyamory, yiffing, felching, snoodling, clear plating, daisy chaining—perhaps you know how that story goes.

Along similar lines, a highly open single male may think all too open-mindedly: "Social taboos notwithstanding, it might be fun to buy a RealDoll for sex play. They cost $6,500, but they are life-size, life weight, and realistic looking, with silicone flesh, synthetic hair, and a fully articulated posable steel endoskeleton. Their facial appearance and body form can also be customized to resemble that of my ex-girlfriend." However, any revelation that one owns a RealDoll hanging from its neck bolt in one's closet may make any future girlfriends much less likely. Too much openness about one's sexuality, if manifest through such consumer purchases, may backfire.

Highly open consumers can be highly profitable, because they can be highly gullible. For example, the more than averagely open constitute the main market for complementary and alternative medicine. Without them, there would be no market for:

- auricular point therapy (acupuncture of the outer ear)
- Bach flower essences (ingestible wildflower essences)
- colonic irrigation (having gallons of warm water pumped through one's colon via anal tubing)
- dolphin therapy (emotional healing through "energy transfers" from contact with dolphins)
- Gerson therapy (drinking vast amounts of fruit juice, plus coffee enemas)

Lest these sound outlandish, bear in mind that there is little more evidence for the efficacy of homeopathy, feng shui, and traditional Chinese medicine, or other goods and services based on magical thinking. All such products are dangerous insofar as they distract people with real illnesses from seeking evidence-based therapies that have actually been demonstrated to work.

## Openness, Novelty, and Fashion

Low-openness consumers remain content with traditional products that have traditional features and designs. They're a tough sell, because they don't value novelty for its own sake. They are happy wearing uniforms instead of fashions: school uniforms, sports uniforms, military uniforms, business suits, golfing wear. They get haircuts instead of hairstyles. In response to marketing sizzle, they're always asking "Where's the beef?" Their instinctive response to the game of runaway consumerism is to stop playing and go boil some potatoes.

High-openness consumers are also much more profitable because they are early adopters and fashion followers. They use the novelty of goods and services to display their openness. Their openness is crucial to consumerism, because it drives novelty seeking, fads, and fashions. This was the great insight of Brooks Stevens, who developed the concept of planned obsolescence after World War II. He noted that businesses can make a lot more money if they continually introduce new product innovations that exploit some consumers' desires to "own something a little newer and a little better, a little sooner than necessary." The products he designed—Lawn-Boy lawn mowers, Evinrude outboard motors, Harley-Davidson Hydra Glide motorcycles—touted minor technical improvements as status necessities for the modern, progress-oriented consumer of the 1950s. They appealed to the highly open: the young, hip, and urban.

Planned obsolescence was central to the golden age of industrial design, from the 1930s through the 1970s. Every year brought new products with avant-garde styling that rendered last year's model conspicuously old-fashioned. As Vance Packard observed, it is impossible to make genuine technical innovations in established products every year, so their planned obsolescence must be focused on design, not functionality. Businesses must seek "planned obsolescence of desirability" (the product is thrown out and replaced because it is no longer fashionable), rather than "planned obsolescence of utility" (the product is thrown out because it no longer works). Stevens's insight spread quickly through all domains of industrial design: cars, houses, clothing, furniture. Business learned that consumers bought things vastly more often if they were convinced that product novelty as an openness indicator trumped product utility and reliability as conscientiousness indicators. So, throughout these decades of dizzying aesthetic innovation, the highly open were highly excited, and the less open were highly confused.

# 13

## Conscientiousness

CONSCIENTIOUSNESS IS the Big Five personality trait that includes such characteristics as integrity, reliability, predictability, consistency, and punctuality. It predicts respect for social norms and responsibilities, and the likelihood of fulfilling promises and contracts. A century ago, people would have called it character, principle, honor, or moral fiber. It is, primarily, the inhibitory self-control imposed by the frontal lobes on the more impulsive, short-term, selfish instincts of the limbic system. Nowadays marketers appeal to impulsive youth by framing older-adult conscientiousness as rigidity, archaism, inhibition, and uncoolness. In a sense, they are right, for there is a trade-off. Conscientiousness is lower on average in juveniles, and as it matures slowly with age, it tends to inhibit spontaneity, fun, and romance—which is to say, it inhibits the short-term mating that tends to maximize reproductive success among young males and that tends to leave young females pregnant and single.

Traditionally, low conscientiousness went by many names: unreliable, undependable, impulsive, flaky; criminal, bum, malcontent, ne'er-do-well, revolutionary. Since powerful elites prefer reliable slaves, peasants, and workers, the general thrust of civilization is to domesticate the young, wild, and impulsive into the mature, conscientious, and hardworking. The colonialist slave owner increases the slave's conscientiousness through whipping to "get his head right." The industrial-era schoolteacher increases the working-class student's conscientiousness through demanding regular attendance, recitation, and homework. The elite traditionally socializes its own youth toward higher conscientiousness through boarding school, military service, and monogamy.

In several respects, conscientiousness is an unusual personality trait. Because hunter-gatherer life did not require as much planning and memory for debts and duties as life in larger-scale societies with more complex divisions of labor, conscientiousness may have evolved to higher average levels only recently, and perhaps to a greater degree in some populations than others. Only with the rise of activities like agriculture and animal herding would our ancestors have needed the sort of anxious obsessiveness and future-mindedness that characterize the highly conscientious. Only in the past ten thousand years did our ancestors prosper by continually asking themselves: Have I plowed enough yet? Did I sow the seeds early enough? Is one of the lambs missing? Did my cousin pay me for those olives? Am I teaching my children the skills they will need in twenty years? Thus were born the sleepless predawn ruminations of the middle-aged conscientious.

For most of adulthood, then, people strive to present a public facade of high conscientiousness—the extreme of the trait dimension most often favored by educators, employers, co-workers, doctors, tax collectors, loan sharks, spouses, children, parents, friends, crops, and domesticated animals. The modern consumerist market obliges by providing a plethora of goods, services, and activities that can function as conscientiousness indicators.

## High-Maintenance Products

Traditional economics assumes that consumers try to maximize their ease and leisure. People should accordingly prefer products that require minimal care, maintenance, and repair. Yet such products make very poor conscientiousness indicators. Much better to buy perishable pet fish, stainable light-colored clothes, furniture with dust-catching ornamentation, cars with glossy paint jobs, and expansive, water-thirsty lawns. Why? Because those low in conscientiousness will not maintain them properly. They will not personally do the nitrate testing, clothes washing, furniture dusting, car washing, and lawn watering. They will forget, or their servants will forget, and the appearance of their possessions will suffer.

This principle—that some people actually prefer products that require some regular maintenance, so they can show off their conscientiousness—often goes unnoticed by economists and unexploited by marketers. Like so much of costly signaling theory, it doesn't make rational sense at first glance. True, any given consumer would prefer to minimize the time and energy he must invest in any given personality display, including any conscientiousness display: he will seek instead shortcuts and time-savers. But at the aggregate level of social signaling, products that become too easy to maintain will lose their value as conscientiousness indicators. They will lose status and reputability. They will be bought and displayed less often by those keen to appear conscientious. For example, solid oak flooring remains more reputable than faux-wood plastic laminate flooring, in part because the oak requires daily spill wiping, monthly mopping with Murphy Oil Soap, and periodic refinishing with power sanders that make so much noise and vibration that all your tropical fish die from stress. (Again, there are ubiquitous trade-offs between different signaling tactics.) Also, as new technologies and design innovations have made it easier to clean and maintain each square foot of household, we keep increasing the total square footage of each house, to maintain a reputably onerous total quantity of housework.

This drive to acquire conscientiousness indicators results in a sort of "conservation of maintenance" principle in most product categories. The modern American megakitchen is a prime example. A major kitchen remodeling job now costs more than $50,000 on average, largely to cover the expense of exotic countertop materials, custom cabinetry, and high-end appliances. The two-career couples who can afford such kitchens rarely have time to use them for cooking, but they wish to demonstrate that they are the kind of conscientious dinner hosts and parents who could, in principle, bake homemade cookies if the need arose. The less food people prepare themselves, the more space and money they tend to devote to displaying their potential capacities for food preparation. My maternal grandmother, Virginia Baker, prepared three meals a day for her twelve children in a seven-by-fourteen-foot kitchen that contained just one four-burner range and a sturdy mixer—no food-preparation island, food processor, disposal, or dishwasher. Today, many suburban families display enough food-prep appliances to feed an army bivouacked in a zoo. Each such appliance reveals the conservation of maintenance principle in its own design. As soon as self-cleaning ovens and easy-wipe burner surfaces were invented, range makers switched from easy-to-clean white porcelain exteriors to hard-to-clean stainless-steel exteriors that show every fingerprint and dog lick. This was not a patriarchal conspiracy to keep housewives busy; it resulted from an intuition among kitchen-proud suburbanites that there is no glory, no conscientiousness-indicating power, in a zero-maintenance range. Far better to get a "commercial" range designed to require nightly cleaning by restaurant staff.

Conscientiousness can be displayed not only through products that require high maintenance, but through those that are conspicuously fragile. Any item that is prone to irreversible entropic catastrophes—such as breakage—will do, because the less conscientious take less care and break more things. A vintage glass Christmas ornament that has been in one's family for four generations testifies to a century of conscientious tree decorating, tree dismantling, and ornament storage. The same message of cautious stewardship can be conveyed by a house full of rickety-yet-intact antique furniture, a closetful of delicate-yet-untorn lingerie, or, as we shall see next, a menagerie of finicky-yet-healthy pets.

## Pets as Conscientiousness Indicators

Chattels that require regular care make excellent conscientiousness indicators, and the pet industry is a prime source of them. Consider the home aquarium for tropical fish. To maintain a modest twenty-gallon freshwater aquarium for a few gouramis, rasboras, and tetras, one must adhere to a tedious maintenance regimen: Daily fish feeding. Weekly partial water changes with awkward gravel-vacuuming tubes and buckets. Weekly water tests for ammonia, nitrite, nitrate, and pH levels. Monthly algae scraping and filter changing. And yet, unlike dogs or people, the fish do not complain if you fail them. They simply die, and you may guiltily join the Facebook group Tropical Fish Loss: A Support Group. Thus, if you still have any fish alive six months after purchasing them, you have demonstrated a high level of conscientiousness. You have kept the fishes' needs in your consciousness, despite their humble size and silent ways, and despite life's manifold distractions. This bodes well for your conscientiousness as a friend, spouse, and parent. Indeed, a single young man with no houseplants and no pets is rightly viewed as a poor prospect by young women seeking Mr. Right. His conscientiousness is either untested (if he never owned plants or pets) or it failed the test (if they died from neglect).

Dogs are even more demanding, hence more reputable as pets. Different breeds allow their owners to demonstrate different forms of obsessiveness. The eerily intelligent breeds (border collie, German shepherd, standard poodle) require regular social interaction and cognitively demanding exercise, or they turn neurotic. The least intelligent breeds (Shih Tzu, bulldog, Afghan hound) require constant vigilance, or they will get themselves killed in some Darwin Awards–style incident. Many small decorative breeds (Maltese, Pekingese, papillon, Yorkshire terrier) require intensive daily grooming and an ever-cautious tread, or they will become bedraggled and two-dimensional. Many aggressive breeds (Rhodesian ridgebacks, Rottweilers, Doberman pinschers) require extensive obedience training, or they will kill neighboring children, which leads to lawsuits, which can dramatically reduce one's home equity. (Rhodesian ridgebacks, bred to kill African lions, can also be hard on neighboring housecats.) Almost every dog breed has some idiosyncratic high-maintenance feature that makes it an effective means for displaying conscientiousness, and all require regular feeding, watering, walking, vet care, and diligent physical restraint by leashes and fences. Hence the social and sexual popularity of single people who can be seen walking dogs that are conspicuously well fed, well-groomed, well trained, nonneurotic, and nondead. (If you find dog care too easy, try owning an African Grey parrot.)

Where genuine organisms are too expensive and space consuming to keep, people buy artificial analogues to function as conscientiousness indicators. The Japanese are especially good at inventing such proxies. The original Tamagotchi (electronic pocket pets) popular with schoolgirls have proliferated into Micro-Babies, Neopets, Nano Pals, Puka-Puka Water Angels, and Zudes. Most require an absurdly high level of maintenance through the clicking of various buttons every few hours. Yet that is precisely the point: their maddening propensity to die if not regularly fed and petted means that one's Tamagotchi can live through a month only if one is a conscientious owner.

There may have been a positive-feedback loop between the domestication of other species and the evolution of human conscientiousness. New research suggests that humans have been shaping the evolution of plants and animals for at least fifty thousand years through various "pre-domestication" processes, and that domestication proper began at least eleven thousand years ago. The cognitive and emotional demands of tending to other species may have shaped the evolution of conscientiousness as a personality trait, driving the human average higher than that found in other social primates. Most species that do "ecosystem engineering" and "niche construction"—creating their own ecological niches by shaping the environment—seem to evolve higher conscientiousness. Beavers raise successful families by building dams that create small lakes where none existed before, and they show quite a work ethic while doing so. The same selection pressures that favored higher conscientiousness may have set the stage for the great advances in human domestication around ten thousand years ago: dogs (for protection and hunting), goats, sheep, cattle, bottle gourds (for containers), figs, bananas, wheat, rice, maize, millet, sunflowers, squash, manioc, and arrowroot. Each demands a different kind of care, a learned set of cross-species empathies that require a foundation of conscientiousness. Such conscientiousness has been well worth advertising to potential mates and friends for many millennia. If we can't display conscientiousness through pride in a herd of healthy goats, we buy Puka-Puka Water Angels instead. As robots become more popular, expect consumers to prefer those that require some degree of care and maintenance—some credible conscientiousness displays—over those that are entirely autonomous. This is why empathy-eliciting toy robots such as Furby, Roboraptor, AIBO, and Pleo sell better than solar-powered robot lawn mowers.

## Collecting

High conscientiousness shades over into obsessive-compulsive behavior. Among consumers, this often takes the form of collecting—acquiring large numbers of products within a particular category, plus the expertise to discuss them intelligently with like-minded collectors. Collections "express one's identity"—as William James wrote in 1890, "A man's Self is the sum total of all that he can call his." However, the act of collecting does not express all the Central Six traits equally. A high, even toxic, level of conscientiousness is foremost in the resourceful acquisition, ordered display, and habitual care of one's collection. Runaway collecting is the apotheosis of runaway consumerism. With the eBay "collectibles" section, everyone can now participate in hoarding and fetishizing objects across most imaginable categories: firefighting memorabilia, Desert Storm militaria, Disneyana, Pez dispensers, snow globes, *Buffy the Vampire Slayer* figurines, nautical decor, replica movie swords, vintage pet-food ads, vintage flue covers, and "risqué" postcards.

Yet one person's obsessive hoarding is another person's rational investment strategy, or his acquisition of tools and materials required for some enthusiasm. I contemplate the twenty-seven hundred pairs of shoes that Imelda Marcos collected in the Malacañang Palace, and feel moral outrage for the poor of the Philippines. I contemplate the comparable number of used books that I have bought, and feel like a dutiful researcher. We can all rationalize our cabinets of curiosities.

## Personal Grooming

The "personal-care sector" or beauty industry is often misunderstood as just a set of products for enhancing physical attractiveness. However, many of its facets make more sense if viewed as conscientiousness indicators.

Consider what we do with our head hair. Humans are unusual among mammals in sporting head hair that grows continuously for about five years, before dropping out. Left uncut, it grows about waist long. Genetic evidence suggests that this innovation evolved about 240,000 years ago. Some have argued that continuously growing head hair evolved as a reliable signal of one's ability to groom oneself and to solicit grooming from others. Contrary to the stereotype of messy-headed cavemen promoted by early museum depictions, it is likely that ancestral humans, like all self-respecting mammals, groomed their hair avidly into diverse styles.

Fashionable hairstyles vary dramatically across history and culture, but they tend to have one common denominator: they require intensive maintenance, whether through regular shaving, cutting, trimming, washing, conditioning, styling, braiding, teasing, gelling, dyeing, powdering, or decorating. The Marine Corps drill sergeant must get his hair cut once a week to maintain the close-shaven macho look. The long-tressed cheerleader must condition, brush, and style her hair daily. The male business executive must shave his face every morning, or risk being mistaken for one of the more "creative types" in marketing. Stubble is acceptable only for backroom financial analyst interns working eighty-hour weeks at investment banks, or CIA agents returning to the office from fieldwork in third-world countries. Likewise, the female business executive must shave her legs every few days, to reassure male co-workers that she submits to the professional-yet-feminine gender role and harbors no secret tendencies toward lesbianism or socialism. The matted dreadlocks that naturally form if human head hair is neglected for months are usually worn only by people who have been ostracized, or who have opted out from mainstream social responsibilities: drunks, stoners, street people, schizophrenics, the shipwrecked. By contrast, well-kept dreadlocks (think mid-1990s Ani DiFranco) can function as indicators of high conscientiousness at a different level: their wearers conspicuously avoid hairstyles that require regular use of alkyl sulfates in shampoos and high volumes of shower water. Grooming fashions and products come and go, because there are so very many ways for humans to impose onerous grooming regimens on themselves. Yet beneath all the variation across time, place, and subculture, the instinct to advertise one's conscientiousness motivates much of this personal care.

## The Unused Exercise Machine

The home fitness industry comprises one of the most profligate wastes of machinery in human history. We pay thousands of dollars for finely crafted running machines, elliptical trainers, and weights. Some of us even buy the Endless Pool, an eight-by-fifteen-foot freestanding "treadmill for swimmers" that produces an adjustable head current, for just $20,400 plus accessories.

These machines were typically purchased with the firm self-improvement convictions of a New Year's resolution or a recent divorce: from now on, we will exercise thrice weekly, lose thirty pounds, and maintain a lean, healthy form well into retirement. We use the machines with jiggly-lardish enthusiasm for several weeks, prattling to our friends and co-workers about feeling rejuvenated. But gradually, disruptions occur. We get a cold, take a trip, suffer a knee injury. Excuses for sloth arise. The frequency of exercise becomes irregular, then lower, then zero. We watch *American Gladiator* (vicarious exercise, at least) while finishing the pint of Cherry Garcia ice cream, and imagine that the elliptical trainer is gazing at us with self-righteous contempt from across the living room. The shame is too great. We move it to the garage or basement, hide it under a tarp, and think, *Someday, when I have more time . . .*

All experienced fitness machine salespeople are well aware that this is the fate of most of their products. What they are really selling consumers is the delusion that the sunk costs of buying the machines will force them to exercise conscientiously. (The consumers know that they could have already been jogging for months around their neighborhood parks in their old running shoes, but they also know that their access to the parks and shoes has not, empirically, been sufficient to induce regular exercise.) So, the consumer thinks: *"If I invest $3,900 in this PreCor EFX5.33 elliptical trainer, it will (1) call forth regular aerobic activity from my flawed and unworthy body, through the techno-fetishistic magic of its build quality, and (2) save me money in the long run by reducing medical expenses."* The salesperson meanwhile thinks: *"20 percent commission!"* and the manufacturer thinks: *"We can safely offer a ten-year warranty, because the average machine only gets used seventeen times in the first two months after purchase."* Everybody's happy, except for most consumers, and they don't complain because they think it's all their fault that they're failing to use the machine. The few conscientious consumers who do use the equipment regularly enjoy many benefits: efficient muscle building and fat burning through the low perceived exertion of the PreCor's smooth elliptical movement; a lean body that elicits lust and respect; a self-satisfied glow of moral superiority.

Thus, home fitness machines make excellent conscientiousness indicators, as they can increase fitness only when used by the highly conscientious; for everyone else, they just gather dust. The result is that, given modern sedentary jobs and overlarge food portions, the only people who can stay in shape throughout their twenties and thirties are the highly conscientious. Everyone else bloats. So, under modern conditions, a lean body testifies not just to the person's heritable body type, but to their heritable conscientiousness, and hence their maturity and reliability as a friend, mate, or co-worker. We infer that if their capacity for guilt and foresight can drive them to regular exercise, it might protect us from being exploited or abandoned by them.

This comfortable signaling system is facing a challenge from "exergaming"—computer games that combine on-screen action with physical exercise. The arcade game Dance Dance Revolution (invented in 1998 by Konami) requires players to activate electronic floor sensors with their feet in fast, complex patterns corresponding to dance steps. Similarly, in MoCap Boxing (2001), players shadow-box against an on-screen opponent, with their moves registered through infra-red motion capture ("mocap") sensors. The Nintendo Wii system brings such exer-gaming into the home. Its remote, which includes accelerometers and an infrared detection system for sensing 3-D location, allows users to play more than five hundred games based on physical activities, such as tennis, baseball, cheerleading, kart racing, cake making, and bal-loon popping, not to mention more speculative possibilities such as chicken shooting, elf bowling, ghostbusting, and Godzilla fighting. These three game systems are not only aerobically demanding but fun to the point of addictiveness and mass popularity, thus threatening to undercut the traditional correlation between conscientiousness and regular use of exercise machines.

## The Credit Rating

In our credit-based economy, a high credit rating functions as a sort of conscientiousness meta-indicator. It is not directly accessible to or perceivable by observers, but it is the basis of so much of our purchas-ing power that a middle-class lifestyle is almost impossible without it. Without good credit, most people cannot acquire a car or house. Thus, individuals who own a decent car and house are indirectly demonstrat-ing their reputable credit score, and hence their conscientiousness.

At least in the United States and United Kingdom, FICO credit ratings from Equifax, Experian, and TransUnion can range from about 300 (very bad) to about 850 (very good). Consider the factors that con-tribute to a high credit rating: Having a longer credit history (that is, being older). Paying bills on time. Sending small amounts of money very regularly to credit card companies, so their interest profits are maximized. Not defaulting on loans or going bankrupt. Living at the same address for many months. Having some debt, but not too much in proportion to one's income. Checking one's credit report every six months for errors (which often arise), and demonstrating the patience to call the credit scoring agency to fix the error. Not having a lot of recent credit score inquiries. This all sounds very straightforward to the highly conscientious. But to average folks, it really is very hard to achieve, which is why the average credit score is around 680—well below the 720+ level for getting the lowest interest rates on credit cards and mortgages.

Now, consider all the fun, impulsive, wild, crazy behaviors that can nuke one's credit score. College students, unfamiliar with the concept that loans must be repaid, tend to incur large debts. Alcoholism and drug addiction tend to cost a lot, create excess debt, result in eviction (frequent changes of address), and interfere with regular bill paying. Several mental disorders (bipolar disorder, borderline personality disorder, attention-deficit disorder, psychopathy) undermine credit scores by driving impulsive behaviors—shopping sprees, sudden changes of job and residence, drug binges and suicidal gestures that incur high hospital bills, reckless driving that incurs high car-repair bills, sexual affairs that require romantic gift-giving, jail terms that reduce income. It is practically impossible for a man with mania, or a woman with borderline, to maintain an above-average credit score. (On the other hand, a touch of obsessive-compulsive disorder, directed toward one's financial behavior, can increase both one's credit score and one's reliability as a social or sexual partner.). In each case—youth, drug addiction, mental disorder—the same impulsiveness that would interfere with having reliable social or sexual relationships also undermines one's credit score. This is why potential friends and mates have incentives to pay attention to one's credit score, albeit indirectly, as it is manifest through one's ability to acquire costly, credit-demanding products.

Strangely, personality psychologists seem not to have investigated the correlation between conscientiousness (as measured by standard personality tests) and FICO credit ratings (as calculated by the rating agencies). Presumably confidentiality issues are involved, but there are also political ones. Personality psychologists know that conscientiousness is a basic, stable, heritable trait. If they showed that it strongly predicts credit ratings, then the credit ratings would start looking less like measures of mystical willpower, and more like measures of brain function. Conversely, variation in average credit ratings across sex, age, race, ethnicity, religion, and social class would raise awkward questions about differences in average conscientiousness across these groups. Thus, it is ideologically useful to ignore the causal links between genes, brains, conscientiousness, credit scores, spending ability, and conspicuous consumption.

## Formal Education and Employment

Pets, lawns, hairstyles, and credit scores are all very well, but so far we have ignored the alpha and omega of conscientiousness indicators: having a good education and a reputable job. These, like the credit score, function as meta-indicators. We cannot buy a genuine education or a stable employment history directly, but it is very hard to achieve either without decent conscientiousness. Thus, school and work, through the regular income they yield, allow the purchase of many other products as indirect conscientiousness indicators.

As all factory foremen and parole officers know, it is extremely difficult for someone of low conscientiousness to hold down a regular job in the modern economy. Woody Allen observed that 90 percent of success is just showing up. The unconscientious, in contrast, show up late for work or do not show up at all. When they do show up, they forget what they're supposed to be doing, make costly errors, don't follow safety protocols, endanger their co-workers, neglect their customers, and annoy their bosses. Lacking the conscience and social inhibitions of other workers, they tend to steal tools and supplies, wreck company vehicles, and take unscheduled holidays. In short, they are more trouble than they are worth. A person of limited intelligence but high conscientiousness can make a valuable employee; a person of higher intelligence but very low conscientiousness is almost unemployable.

Educated professionals (like most readers of this book) are largely insulated from such low-conscientiousness fiascos on the job, because their support personnel are prescreened for a decent level of conscientiousness. It is when they try to find good babysitters, gardeners, car mechanics, and building contractors that the true range of conscientiousness in the general population becomes manifest. For example, the most difficult thing about remodeling a house is dealing with general contractors who do not return phone calls, specialist contractors who do not show up, and suppliers who do not deliver. After six months of such behavior, most homeowners have a little less respect for people who look fun and cool, and a little more respect for people who are consistent and diligent.

There are class differences in how one's work life displays one's conscientiousness. Working-class hourly employees who must punch time clocks show their conscientiousness through reliability of attendance, regularity of work hours, and quantified consistency of job performance (for example, a low rate of assembly-line errors or forgotten entrée orders). Middle-class salaried workers show conscientiousness through having completed some consistency-demanding formal education and by completing specific tasks on time for their line manager. The self-employed, the small-business owners, and the creative class (writers, artists, marketing consultants) have an especially acute challenge: demonstrating conscientiousness through long periods of diligent work, unreinforced by bosses, social pressures, or sharp deadlines. The freelance science journalist, corner-shop grocer, and prolific genre novelist have this in common: if they are making a viable living doing their job, they must have high conscientiousness. In fact, at the highest levels of self-directed achievement, workers tend to seek out minimally structured jobs almost as a self-handicapping way to show extreme diligence. The tenured professor has minimal external incentives to remain an interesting teacher and productive researcher, so if she does, it speaks especially clearly of her intrinsic dutifulness.

Usually the amount of oversight that a job entails is assumed to be a logical product of the job's intrinsic nature, rather than the worker's desire to display his diligence optimally. Yet many professionals seem to seek out understructured jobs, and join professional guilds such as bar associations and medical associations that lobby for minimal oversight. They do this not because they are unconscientious slackers, but because they seek the extra challenge of remaining productive despite the temptations to loaf. Thus, the highest-status professions are those in which sustained conscientiousness is required for long-term career success, but in which there are minimal sticks, carrots, and bosses to motivate short-term performance. This applies even to celebrity roles that look easy to outsiders. Most successful actors must endure hundreds of failed auditions before making their name. Professional athletes must spend years training from childhood onward. Aspiring U.S. presidents must campaign for two solid years before elections.

School, work, and credit—three pillars of consumer capitalism—are also, not coincidentally, the most reliable and conspicuous indicators of conscientiousness. All other consumer purchasing depends on these three pillars, so they are fundamental to conspicuous consumption.

# 14

# Agreeableness

AGREEABLENESS IS NOT just one of the Big Five personality traits. Construed more broadly as a personal capacity for empathy, kindness, and benevolence, and as a desire for egalitarianism and social justice, agreeableness is at the heart of human altruism and social progressivism. It is the rare product of natural selection and sexual selection that makes our species seem to transcend the otherwise selfish imperatives of the evolutionary process. It is our last, best hope for the salvation of our species, but also our most persistent source of hypocrisy and runaway self-righteousness. And we advertise agreeableness in such diverse and subtle ways, through such complex and intangible products, that we often do not realize we are signaling it at all.

## The Agreeable Economy

Economies are driven by this trait at several levels: by high-agreeableness consumers striving to display their kindness and generosity, and by low-agreeableness consumers striving to display their assertiveness and dominance.

At the macro level, ritualized occasions for gift giving to mates, friends, or relatives promote a huge proportion of retail sales. Most retailers would go bankrupt quickly if their culture did not celebrate some subset of calendar holidays (Christmas, Boxing Day, New Year's Day, Fasching, Valentine's Day, Easter, Vesak, Mother's Day, Father's Day, summer holidays, Ramadan, Labor Day, Oktoberfest, Diwali, Halloween, Thanksgiving, Chanukah, and so on), family remembrances (birthdays, anniversaries), and rites of passage (bar mitzvahs, graduations, marriages, honeymoons, baby showers, retirements, funerals). These all depend on the agreeable signaling their agreeableness by giving gifts or feasts to the grateful.

At the micro level, many products thrive because they are associated with agreeable personalities and activities. Since the 1930s diamond engagement rings have been the premier symbol of romantically honorable intentions and likely spousal agreeableness. Early twentieth-century women faced a problem: prosecution of men for financial damages following breach of promise was declining. It was becoming all too common to be seduced by a psychopath promising marriage and then abandoned after he availed himself of one's virginity during the engagement. Into this reliable-signaling gap jumped De Beers with the diamond ring, heavily promoted with the slogan "A diamond is forever." Diamond marketers recommended that women ask men to spend two months' salary (or about a year's disposable income) on a ring, as a sign of the seriousness of their commitment. Ever since, engagement rings have dominated the demand for diamonds larger than one carat. Every time men found a cheaper source of diamonds, women demanded larger ones to maintain the signaling reliability. Now, aspiring grooms are stampeding from retail jewelry stores to online stores such as Blue Nile that charge 40 percent less for the same ring. Nonetheless, their fiancées, who typically pick their favored ring design from Blue Nile themselves, still expect the man to follow the two-months'-salary rule (yielding an average engagement ring price of $6,400). The online savings is converted into a larger diamond, not into a smaller cost, because a smaller cost would represent a less-reliable signal of agreeableness.

Many services are also marketed as amplifiers of agreeableness. These usually teach "etiquette," that is, how to emulate the tacit social norms of the local ruling class. Such norms usually require practicing superhuman levels of patience, discretion, generosity, and sympathy; the implicit goal is to demonstrate that one's prefrontal cortex can maintain tight inhibitory control over selfish or impulsive behaviors. It has always been crucial for ruling-class youth to acquire such conspicuous agreeableness indicators, so they can evaluate one another's capacities for peaceful and efficient cooperation, which is vital to the smooth operation of the various conspiracies that secure their wealth and power, such as feudal aristocracies, organized religions, trade guilds, parliaments, and media conglomerates. Traditionally, Europeans bought etiquette training for male offspring at boarding schools (such as Eton or Sandhurst) and universities (such as Oxford or Cambridge), and for female offspring in Italian convents or Swiss finishing schools (such as Surval Mont-Fleuri or Institut Villa Pierrefeu).

## Indicators of Agreeableness Versus Aggressiveness

As all marketers know, there is gold in the eighteen-to-thirty-four-year-old male demographic group. Young adult males invest far more time, energy, and money in mating effort than in parenting effort, and mating effort yields the most profitable products. Depending on their current mating tactics and rivals, young males switch often between displaying low agreeableness and displaying high agreeableness. When a young male is trying to impress a female initially with his general mate value, social dominance, maturity, and manliness, or trying to deter male sexual rivals from hitting on the chosen female, he must display low agreeableness: a risk-taking, violent assertiveness bordering on psychopathy. On the other hand, once a relationship is established, the young male must display high agreeableness to signal his potential as a long-term mate: romantic thoughtfulness to the women, gentleness to children and animals, concern for the environment, social justice, and family values. This may be why popular music played by young males alternates shamelessly between fast, heavy, aggressive anthems (to excite ovulating groupies and intimidate male rivals), and slow, soft, romantic ballads (to reassure the highest-quality female girlfriends that the males will be forever chaste and loving).

This may also be why males tend to switch from assertiveness displays in young adulthood to agreeableness displays in middle age. For example, teens and college boys tend to want big, fast, loud vehicles. If they like motorcycles, they will fantasize about wearing a Kevlar-

and-titanium crash jacket while riding a dyspeptically noisy Harley-Davidson Fat Boy ($17,000). Or, better yet, the Boss Hoss BHC-3 ($40,500), a thirteen-hundred-pound motorcycle made in Dyersburg, Tennessee, with an 8.2 liter, 502 horsepower V-8 engine. They will not seek dates on Caringsingles.com ("a new Internet dating site serving people who care about the world, the earth, and living life fully"), but in the gothic promiscuity maze of MySpace.

This drive to assertiveness display is especially obvious for the single most powerful, expensive, and dangerous product that young males tend to buy: the automobile. Cars have always been advertised as symbols of sexual potency and conquest, but the rhetoric has become more extreme with the SUVs of the 1990s and the muscle cars of the 2000s. Examples from ads for the Subaru WRX: "Remember, this thing's loaded"; "You have to muscle your way to the front of the pack." The latent prison-gang-rape aggressiveness of many American SUV model names becomes all too apparent if one prefixes the word "anal" in front of them, which yields the: Anal Armada, Anal Ascender, Anal Commander, Anal Endeavor, Anal Expedition, Anal Explorer, Anal Hummer, Anal Pathfinder, Anal Torrent, Anal Trailblazer, Anal Tribute, and Anal Wrangler. As jingo-fascist rage replaces social civility, I wouldn't be surprised to see even more aggressively branded SUVs sold in the United States soon, maybe a Dodge Daisy Cutter, Ford FUBAR, or Buick Water-Boarder. Also look for arms manufacturers to develop automobile brand extensions: perhaps a Glock roadster or Kalashnikov coupe.

What does not work is to mix features that signal agreeableness (an environmentally friendly hybrid drive) with features that signal disagreeableness (huge size and intimidating front end), as in the 2008 Chevrolet Tahoe Hybrid, "America's first full-size hybrid SUV." It simply doesn't make sense to most onlookers that an eco-consciously agreeable person would drive a full-size SUV that weighs fifty-eight hundred pounds and seats eight—or that a dominant Chevy driver would pay the $14,000 price premium for a hybrid drive that only improves mileage from 16 mpg to 20 mpg.

Twenty years later, after their testosterone levels decline to reduce their mating effort, the same males who sought the Boss Hoss or Glock roadster will mostly be settled down, and understand that their wives and kids value agreeableness more than aggressiveness. They will find themselves behind the wheel of a large Toyota Camry hybrid. They will consult the American Anti-Vivisection Society *Guide to Compassionate Shopping*. They will wonder whether it is more important to attend that evening's PTA meeting or check the R-value of their roof insulation or read that new bestseller, *The Subordinate Husband's Guide to Giving Sensual Massages with No Sexual Expectations*.

## Displaying Agreeableness Through Conformity

Highly agreeable people want to get along with everyone, so they tend to be conformists, whether with respect to peer-group opinions, fashions, or product choices. Conversely, anticonformity can signal dominance, assertiveness, and low agreeableness.

To test the idea that people use conformity strategically to signal agreeableness, Vladas Griskevicius and his colleagues ran another "mating prime" study. They expected a sex difference, because women have a stronger preference than men do for mates who display assertiveness, dominance, leadership, and risk taking. So, mating-primed males may try to display these lower-agreeableness traits through conspicuous anticonformity—by resisting and rebelling against peer influence. On the other hand, mating-primed females may try to display their higher-agreeableness traits (kindness, empathy, social-networking ability) through conspicuous conformity to peer influence.

Subjects were randomly assigned to one of three priming conditions. In the mating-prime condition, they read a romantically arousing story about being on a vacation with friends, meeting and spending the day with a highly desirable person of the opposite sex, and kissing passionately on a moonlit beach. In the "threat prime," they read a frightening story about an intruder breaking and entering when they were home alone at night. In the "neutral prime," they read a happy story about going to a much-anticipated live music event with a same-sex friend. After experiencing one of these primes, the subjects were shown various artistic images. They were told that all three of their peers gave either positive or negative ratings to each of the images, and then they gave their own ratings. Their level of agreement with the peers indicated their degree of conformity.

As predicted, Griskevicius found that mating-primed men showed less conformity than in the threat or neutral conditions, whereas mating-primed women showed more conformity. These mating-prime effects were modulated in a fascinating way by the direction of the peer evaluations. If all the peers rated a particular artistic image positively, mating-primed men showed neither conformity nor anticonformity; they just followed their previously measured aesthetic tastes. But if all the peers rated a particular artistic image negatively, mating-primed men showed strong anticonformity (and thus higher openness) by rating the image much more positively. However, mating-primed women showed stronger conformity if all their female peers rated the artistic image positively, and neither conformity nor anticonformity if their peers rated the image negatively. It looks as though each sex wants to act "positive" in their aesthetic ratings, but the males prefer to act positive most strongly when all the other males act negative, whereas the females prefer to act positive most strongly when all the other females are also positive. Conformity interacts with positivity in the strategic signaling of this personality trait. (By contrast, the threat prime concerning the home intruder led both sexes equally to show higher conformity in their ratings of the artistic images, as if a self-protection motive were favoring group-mindedness.)

In a follow-up study, Griskevicius discovered a further nuance in human self-presentation: the sex-specific effects of the mating prime on conformity are influenced by whether a person's judgment concerns subjective taste or objective fact. Mating-primed males show especially strong nonconformity when they make subjective judgments about which consumer product they would prefer (a Mercedes or BMW luxury car, a Ferrari or Lamborghini sports car), but they switch to showing very high conformity when they are asked objective knowledge questions (is it more expensive to live in New York or San Francisco? Which airline has more on-time arrivals, Southwest or America West?). So, mating-primed men want to stand out from the crowd when it comes to having distinctive taste, but they rely on peer opinion to avoid factual errors. On the other hand, mating-primed females show strong conformity when making the subjective judgments, but they show neither conformity nor anticonformity when answering the objective questions.

Thus, men seem especially keen to show off their assertiveness and independence through their anti-conformity when they want to impress a woman, as long as the anticonformity doesn't make them look more negative and closed-minded than their rivals, and doesn't lead them to make an embarrassing factual error. Women are keen to show off their agreeableness through conformity when they want to impress a man, especially when they're conforming to a positive, open-minded judgment. At least in these experiments, women were less influenced than men by peer opinion when answering factual questions. So, it would be extremely simplistic for marketers to claim that consumers are generally conformist, or that women are more conformist than men, or that people are more conformist about preferences than facts, or that emotional arousal makes people more conformist. It all depends on the social and emotional context, the traits that people are trying to display, and the traits that other people tend to value.

## Ideology as an Agreeableness Indicator

Young adults of both sexes often devote massive amounts of time, money, and energy to signaling their agreeableness through their ideologies. For example, at Columbia University in 1986, there was a sudden upsurge of conspicuous agreeableness one spring. Hundreds of college students took over the campus administration building and demanded that the university sell off all its stocks in companies that do business in South Africa. (This was in the days of apartheid, when

Nelson Mandela was still in jail and blacks could not vote.) The spontaneity, ardor, and near-unanimity of the student demands for divestment seemed puzzling. Why would mostly white, mostly middle-class North Americans miss classes, risk jail, and occupy a drab office building for two weeks in support of political freedom for poor black strangers living in a country eight thousand miles away? The campus conservative newspaper ran a cartoon depicting the protest as an annual springtime mating ritual, with Dionysian revels punctuated by political sloganeering about this year's arbitrary cause. The cartoon seemed patronizing at first, but later it seemed to contain a grain of truth. Although the protests achieved their political aims, only inefficiently and indirectly, they did promote very efficient mating among young men and women who claimed to share similar political ideologies. Everyone seemed to be dating someone they'd met at the sit-in. In many cases, the ideological commitment was paper-thin, and the protest ended just in time for the students to study for semester exams. Yet the sexual relationships facilitated by the protest lasted for years in some cases.

It seems cynical and dangerous to suggest that loud public displays of one's political ideology function as some sort of courtship ritual designed to attract sexual mates, for it risks trivializing political discourse, just as the conservative cartoon did when lampooning the Columbia antiapartheid protests. The best way to avoid this pitfall is not to ignore the costly signaling logic of human political behavior, but to analyze it seriously and respectfully as a dramatic example of personality display.

Humans are ideological animals. We show strong motivations and capacities to learn, create, recombine, and disseminate value-laden idea systems, often with a righteous contempt for any empirical evidence that would undermine them. Yet it has always seemed hard to envision a survival payoff for conspicuous ideologies that scoff at empirical reality. Fortunately, costly signaling theory does not demand survival payoffs, only social and reproductive payoffs. If a conspicuously displayed ideology correlates reliably with a certain set of personality traits that are socially and sexually desired, then the ideology's empirical truth is irrelevant. Indeed, the most empirically misleading and self-handicapping ideologies might often make the most reliable personality indicators.

The vast majority of people in modern societies have little political power, yet they do have strong political convictions that they broadcast insistently, frequently, and loudly when social conditions are right (political protests, dinner parties, second dates). This behavior is puzzling to economists, who regard any ideological behavior—even voting—as an expenditure of time and energy that has little political benefit for the individual. But if we view the individual benefits of expressing political ideology as usually not political at all, but rather as social and sexual, we can shed light on a number of old puzzles in political psychology. Why do hundreds of questionnaires show that men are, on average, more conservative, more authoritarian, more rights oriented, and less empathy oriented than women? Why do people usually become more conservative as they move from young adulthood to middle age? Why do more men than women run for political office? Why are most ideological revolutions initiated by young single men?

None of these phenomena make sense if political ideology is interpreted as a rational reflection of political self-interest. In political, economic, evolutionary, and psychological terms, everyone has equally strong self-interests, so everyone should engage in equal amounts of ideological behavior, if that behavior functions to advance political self-interest. However, we know from sexual selection theory that not everyone has equally strong reproductive interests. Males have much more to gain from many acts of intercourse with multiple partners than do females, because males can potentially produce offspring by hundreds or thousands of different women, but women can bear only about a dozen offspring in a lifetime. Young males should consequently be especially risk seeking in their reproductive behavior, because they have the most to win and the least to lose from risky courtship behavior (such as becoming a political revolutionary). These predictions are obvious to any sexual selection theorist; less obvious are the ways in which political ideology is used to advertise different aspects of one's personality across the life span.

Adults, especially when young, tend to treat one another's political orientations as proxies for personality traits. Conservatism is read as indicating an ambitious, self-interested personality that will excel at protecting and provisioning a sexual partner. Liberalism is read as indicating a caring, empathetic personality that will excel at child care and relationship building. Given the well-documented, cross-culturally universal sex difference in human mate choice criteria, with men favoring younger, fertile women, and women favoring older, higher-status, richer men, the expression of more-liberal ideologies by women and more conservative ideologies by men is not surprising. Men use political conservatism to (unconsciously) advertise their likely social and economic dominance; women use political liberalism to advertise their nurturing abilities. The shift from liberal youth to conservative middle age reflects a mating-relevant increase in social dominance and earning power, not just a rational shift in one's self-interest.

More subtly, because mating is a social game in which the attractiveness of a behavior depends on how many other people are already producing that behavior, political ideology evolves under the unstable dynamics of social imitation and strategizing, not just as a process of simple optimization, given a particular set of self-interests. This explains why an entire student body at an American university can suddenly act as if it cared deeply about the political fate of a country that it virtually ignored the year before. The consensually accepted way to display agreeableness simply shifted, capriciously and quickly, from one political issue to another. Once a sufficient number of students decided that attitudes toward apartheid were the acid test for whether one's heart was in the right place, it became impossible for anyone else to be apathetic about apartheid.

What can we do to improve society if most people treat political ideas as courtship displays that reveal their advocates' personality traits, rather than as rational suggestions for improving the world? The pragmatic, not to say cynical, solution is to work with the evolved grain of the human mind by recognizing that people respond to policy ideas not just as concerned citizens in a modern polity, but also as hypersocial, status-seeking primates. This view will not surprise political marketers (pollsters, spin doctors, speechwriters), who make their living by exploiting our lust for ideology, but it may surprise social scientists who take a more rationalistic view of human nature. Nonetheless, to understand a great deal of consumer behavior, we have to acknowledge the fundamentally ideological nature of many purchasing decisions, and the way that everyone uses products, in various ways, to advertise his or her personality traits.

## The Religious and Political Service Industries as Personality Indicators

In cosmopolitan societies, even religious and political ideologies are marketed as personality indicators, especially as indicators of agreeableness or assertiveness. The religious services industry today is a trillion-dollar global business; the Catholic Church in the United States alone takes in more than $100 billion in revenue per year. Religion used to be a set of local monopolies segregated by ethnicity, culture, and language, but over time, rituals, laws, and taboos became more elaborate as runaway indicators of high conscientiousness and low openness. New sects began to break off when young worshippers wanted to show off different personality traits, such as openness, extraversion, and agreeableness (for example, revolutionary early Christians versus conservative Pharisees, inclusive Mahayana versus exclusive Hinayana Buddhists). This gradual market segmentation in the religious services industry replaced the regional monopolies (Europe's Catholicism, India's Hinduism) with a proliferation of start-up religions. Consumer choice of religion became guided less by family tradition and more by individual selection of personality display strategies. Thus, you can probably make some valid character inferences about people today who choose to be Quakers (agreeable, intelligent) versus

Satanists (disagreeable, impulsive), or Zen Buddhists (open, stable) versus Orthodox Jews (conservative, conscientious). Recent spoof religions can be even more revealing. If your office cubicle includes a coffee mug sporting the slogan "WWFSMD?" (What would the Flying Spaghetti Monster do?) co-workers can readily infer that you are an ironic, secular, highly open "Pastafarian" who supports evolution and opposes "intelligent design" and school prayer.

The same is true of the political ideology industry. American Democrats and British Labour voters tend to be more agreeable and more open; Republicans and British Conservative voters tend to be more conscientious and less open. Thus, an Obama/Biden bumper sticker in 2008 was not just a content-free injunction to vote Democrat; it was a content-rich display of the driver's social, sexual, and cultural openness. The personality-indicating function of political ideology helps explain why political arguments are usually a waste of time. Trying to convince someone to switch from Green to Libertarian on the basis of rational arguments and empirical evidence is as futile as trying to change someone's inherited personality type by these means. Likewise, regardless of abstract political principles, anxious introverts won't favor legalizing Ecstasy, and chaste wives won't favor legalizing prostitution. Odd positions taken by political parties are often attempts to maintain their personality-indicating power rather than logical expressions of basic moral-political principles. For example, Republican opposition to gay marriage is not derived from the party's 1854 foundational beliefs in Federalism, abolitionism, and free speech, but it is simply consistent with modern Republican identity as the party of high chastity ("family values") and low openness ("conservatism").

## Signaling Failures in Ideology

One of the most frustrating experiences in human life is to adopt an unfashionable new worldview after much evidence-based research, rational consideration of alternatives, and ethical soul-searching, only

to have one's peers misconstrue that worldview as a personality signal that conveys the opposite of one's true traits and intentions. This
is a common experience among evolutionary psychologists, and it
nicely illustrates the way that ideology signals can fail under certain
conditions.

Critics such as Stephen Jay Gould, Steven Rose, and Richard
Lewontin have convinced a substantial portion of the educated public that evolutionary psychology is a pernicious right-wing conspiracy,
with the hidden ideological agenda of reviving biological determinism,
sexism, racism, and elitism. They conflate the worst excesses of 1860s
social Darwinism, 1890s union-busting capitalism, 1930s Nazi eugen-
ics, and 1970s sociobiology with the twenty-first-century science of
human nature.

What the critics fail to explain, however, is why evolutionary psychology has attracted the support of so many socially conscious progressive thinkers, ranging from the animal rights philosopher Peter
Singer to the economist Robert Frank, the archcritic of runaway consumerism. They likewise fail to explain why so many prominent evo-
lutionists (E. O. Wilson, Robert Trivers, John Maynard Smith) have
had strong ties to left-wing politics in their private lives. And they fail
to explain why right-wing American fundamentalists see evolutionary
psychology as an ultraliberal attack on family values and religion.

To determine whether evolutionary psychologists are actually right-
wing conspirators, my colleagues Josh Tybur, Steve Gangestad, and I
conducted an online survey of 168 psychology Ph.D. students (from
six major U.S. universities), 31 of whom self-identified as "adaptationists" (evolutionary psychologists), and 137 of whom identified with
some other theoretical framework. Both groups were about 70 percent
female and about twenty-seven years old on average. Of the 31 evolutionary psychologists, none identified with the Republican Party, and
only 2 identified with the Libertarian Party. None of them voted for
George W. Bush in 2004, and only 1 voted for the Libertarian Michael
Badnarik. (This contrasts with the 30 percent of Americans who iden-

tify with the Republican Party, and the majority of voters who alleg-edly supported George W. Bush in 2004.) On sixteen questions about political attitudes, evolutionary psychologists scored significantly more liberal than conservative on fifteen of them. (They were evenly split on the last item.) For example, of the 31 evolutionary psychol-ogy Ph.D. students, 30 supported gay marriage, 29 supported envi-ronmentalism, 26 supported abortion rights, 25 supported universal health care, 21 opposed cutting the federal income tax, 19 supported marijuana legalization, 17 supported raising the minimum wage, and 17 opposed preemptive military actions against foreign countries. On political issues, evolutionary psychologists were just as liberal as non-evolutionary psychologists (and even more liberal on individual rights and social libertarian issues), and vastly more liberal than the gen-eral U.S. population. Our sample size of 31 evolutionary psychologists is not a high number, but it is a high proportion—roughly one-third to one-half—of all students then enrolled in U.S. evolutionary psy-chology Ph.D. programs. We can be fairly confident that evolutionary psychologists are politically similar to academic psychologists in gen-eral. To the religious right, and even to most Americans, we look like godless, gay-friendly, tree-hugging pacifists. Only the old guard of the 1960s New Left (Gould, Rose, Lewontin) could possibly mistake us for conservatives.

Thus, we have a signaling failure. Holding an evolutionary psy-chology worldview is still perceived by many educated people as an indicator of conservatism, disagreeableness, and selfishness, through a process of guilt by historical association with 1860s social Darwin-ism. Yet empirically, holding an evolutionary psychology worldview is actually an indicator of liberalism, agreeableness, and altruism. Such mismatches between the apparent information conveyed by an ideol-ogy and the ideology's actual correlations with personality traits, are probably common for ideologies that are new or rare. People just don't have enough direct personal experience with the ideology adherents to disconfirm their media-fed prejudices. This explains why dominant ideologies can so easily maintain their monopoly power as signaling systems: they can portray alternative ideologies as signaling undesirable personality traits, and thereby preempt any signaling benefit of switching ideologies. The result is a potent lock-in effect: certain ideologies have certain personality correlates by historical association and common knowledge. So, it is very hard for advocates of a new ideology to gain a beachhead in public consciousness, unless the new ideology wears its personality correlates very conspicuously on its sleeve.

# 15

# The Centrifugal Soul

How can we as individuals live happily in a postconsumerist style that still lets us show off our personal traits, so that we obtain the social and sexual attention we crave? I think it helps first to consider how much of one's "soul"—one's abilities, personality traits, preferences, and values—one is displaying too obsessively, too preemptively, too dizzily, and too publicly. Modern consumers are like children doing spin art: we pour our favorite colors all over a fast-spinning lifestyle, fling pigment in all directions, and hope that some will stick to observers long enough for them to notice our composition. Given sufficient resources, a few of us do succeed in attracting attention, so that we become the human equivalent of a six-foot-diameter Damien Hirst spin painting, such as *Beautiful revolving sphincter, oops brown painting* (2003). Most of us, however, just make a mess. We put too much of ourselves into our product facades, spinning too much mass to our outer edges where we hope it is both publicly visible and instantly lovable. One problem with this strategy is that it leaves too much blank space in the middle, so there's not much of ourselves left for lovers or friends to discover in the longer term. This could be called the centrifugal-soul effect: runaway consumerism leaves us feeling superficial and empty, because we project ourselves outward to observers too promiscuously and desperately. We forget the virtues of restraint, reticence, and dignity. We lose our capacity for self-contained, self-sufficient self-judgment. We end up like a country without external borders or internal tradition. This chapter suggests some ways that we as individuals can overcome this centrifugal-soul effect. Later chapters address some changes in social norms and government policies that could broaden the alternatives to conspicuous consumption.

## The Renunciation Strategy

Historically, those who renounce conspicuous consumption have most often been religious or political zealots, such as Calvinists or Marxists. They usually reject not just runaway consumption, but all forms of self-display and all status seeking, which are seen as mortal sin, bourgeois decadence, or false consciousness. These humble egalitarians have included Hindu ascetics, Buddhist monks, Christian saints, Puritans, and hippies. Of course, human nature always leads such renouncers to construct new status hierarchies of their own, based on costly behavioral displays of conscientiousness, introversion, and emotional stability—and the fortitude to survive starvation, poverty, chastity, and loneliness. Even the sadhus of India, who may sit for years with one arm lifted overhead, visibly atrophying, attain a certain social status—and receive food donations—from their conspicuous rejection of conspicuous consumption.

Modern renouncers are more likely to join the voluntary-simplicity movement by subscribing to *Real Simple* magazine and carrying a *Slow Food USA* tote bag. In either case, the renouncers remain awesomely self-deceived in believing that they have left behind the whole castle of self-display just by escaping the dungeon of runaway consumption. Since this type of self-deception looks naïve and witless to those who understand the evolutionary origins and functions of self-display— including my dear readers, by now—the renunciation strategy itself ends up looking stupid and childish. It speaks highly of the renouncer's conscientiousness and agreeableness, but poorly of their intelligence, experience, and insight.

## Alternatives to Retail

It seems far more self-aware and creative to take a hard, conscious look at one's self-display strategies—to assess their true social and sexual goals, their reliability and efficiency as trait displays, and the many alternatives that are available.

The standard self-display strategy in most developed societies is to seek the highest-paying full-time employment permitted by one's intelligence and personality, and to use the resulting income to buy branded goods and services at full retail price. Weekdays are spent working; evenings and weekends are spent shopping. The purchases must serve a conflicted dual role in signaling. On the one hand, their invidious price must testify to the traits of rational intelligence, conservatism, conscientiousness, and amorality demanded by most high-paying professions and corporations. On the other hand, their style and leisure uses must testify to the opposite traits of emotional warmth, openness, spontaneity, and virtue demanded by most mates and friends. The result is the bourgeois bohemian phenomenon so well described by David Brooks in *Bobos in Paradise* (2001): aspirants must act like obsessive brainiacs at work, and then switch personalities 180 degrees to act like sensual, eco-minded slackers at home. At work, they must embrace Adam Smith's division of labor: specialize, rationalize, delegate, globalize. At home, they must pretend to enjoy making their own pasta. The commute becomes an existential transition zone between one false persona and its opposite.

As a self-display strategy, it is very inefficient to buy new, branded, mass-produced products from stores at the full manufacturer's suggested retail price. The product comes into one's life naked and mute, without any social context, memorable circumstances, or narrative value. Nothing about the purchase says anything about one's traits, except one's ability to afford the purchase. One can't talk about the product as a distinctive object with a unique provenance. One can merely own it, use it, display it, and hope that someone appreciates its wealth display function. Almost every other way of acquiring and displaying human artifacts or experiences sends richer signals about one's personal qualities—though it usually brings less revenue to the retailer and manufacturer. In each case, the product's signaling value is hugely augmented by the human capacity for language—by what one can say about the skills required in making, finding, acquiring, maintaining, and/or repairing it.

Suppose there is a product that you think you want to buy. You've seen the ads; you covet the item—perhaps some new iPea with ten terabytes of memory packed into a tiny green sphere. You anticipate the minor mall adventure: the hunt for the right retail environment playing cohort-appropriate nostalgic pop, the perky submissiveness of sales staff, the quest for the virgin product, the self-restraint you show in resisting frivolous upgrades and accessories, the universe's warm hug of validation when the debit card machine says "Approved," and the masterly fulfillment of getting it home, turned on, and doing one's bidding. The problem is, you've experienced all this hundreds of times before with other products, and millions of other people will experience it with the same product. The retail adventure seems unique in prospect but generic in retrospect. In a week, it won't be worth talking about. So it goes, in the world of mass manufacturing, mass marketing, and mass retailing.

Fortunately, there are alternatives to buying new, branded, mass-produced products at full retail price from anonymous sales staff in unmemorable stores. They are not easier, but they carry much higher signaling value about one's personal traits. For example, whenever you think you want to buy a new physical product, consider these options instead.

## Just Don't Get It

Think for a few days about whether you really need the product at all—especially its conspicuous display features and functions. Consider that talking later about one's conscious decision not to acquire the product may carry more memorable information about your character than acquiring it. Consider whether it really will deliver the social and sexual status boost implied by the advertising, and whether such a status boost is really worth the cost, given the likely disruptions to your ongoing social and sexual relationships. (A sexy new $40,000 convertible, for example, might lead to a $400,000 divorce.) Sometimes, the itch to shop is just the body's way of telling you to stop being so sedentary. Instead of going out to spend money, just use your long-neglected exercise machine, walk the dog, or canoodle with your lover.

Also, for some indulgences, it's worth considering how much you would pay *not* to own the item. For example, Costco sells M&M candies in sixty-four-ounce bags for about $8. I like M&Ms, so that seems like a great impulse purchase if I think I deserve a treat. However, at 142 calories per ounce, that bag contains about 9,000 calories of milk chocolate, which, knowing myself, I would eventually eat. An intensive aerobics class burns only 500 calories an hour, so it would take eighteen hours of aerobic lessons, at $10 per hour, to counteract the fat gain. So, rationally, I should be willing to pay about $180 to the Costco cashier—or my wife, or anyone—to restrain me from buying the $8 bag of M&M's. That's the value of willpower in this case. Similar reasoning can help keep one from acquiring sports sedans with high insurance and maintenance costs, houses with high property taxes and heating costs, mobile-phone packages with high monthly charges, printers with expensive ink cartridges, or degrees from universities that are especially effective at soliciting alumni donations. For many products, the long-term net costs of ownership and consumption far outweigh the short-term benefits.

### Find the One You Already Own

If you're older than age thirty, and the product is a physical object that wasn't just invented, you probably already have one in your garage somewhere. Find it, clean it off, make it work, and use it instead. If you have to repair or upgrade it, so much the better—those are skills you can talk about when you use it in front of others. If your old product lacks certain features of newer products, it may have other compensating charms (familiarity, retro design, proven reliability). Also, the new features can sometimes be retrofitted. If your old V-6 car lacks the acceleration of the new V-8s, just add an aftermarket supercharger for $3,000; don't buy a whole new vehicle. If it lacks the premium leather seats of the new model, you can get all-leather custom seat covers for your old car for about $1,200—two-tone, with piping and your personal logo—from Trimcar.com or other retailers.

## Borrow One from a Friend, Relative, or Neighbor

Most big purchases are rationalized like this: the high initial capital cost will be amortized over many years of use. The product will be used hundreds of times, so the cost-per-use will grow ever smaller, and one's savings will grow ever larger. However, consumers with good memories and minimal self-deception will realize that very few of their purchases ever justified their cost in this way. Because so many products get used a few times and are then forgotten, it is better to borrow them from an acquaintance, use them a few times, and then return them rather than having to store them. Borrowing also builds up social capital, as it promotes reciprocity, trust, and social bonds. If you are conspicuously using an expensive product borrowed from a friend, it testifies to that friend's trust that you will use it carefully and return it promptly—key components of conscientiousness.

## Rent It

Given that you won't use most products nearly as often as you think you will, it often makes sense to rent instead of buy. The sense of ownership is a cognitive illusion, in any case, founded on a denial of habituation, consumer fashion cycles, technical progress, and our own mortality. Buying something outright means, at best, being able to use it until it becomes boring, unfashionable, or obsolete, or until one dies. Personal "ownership" is just a way of renting things from the universe for a human life span, or less. He who dies with the most toys . . . should have bought fewer toys and rented more. The true cost of ownership per product use is simply the price you pay for something when you buy it, minus the price you get when you sell it, give it away, lose it, or throw it away, divided by the number of times you use it. This cost is often amazingly high, and economists are routinely puzzled by the human propensity to overbuy and underrent products that will see only occasional use.

For example, if you truly think that driving a Ferrari F430 will impress a particular person on a particular date, just spend the $1,750 to rent it for a day from Gotham Dream Cars or wherever, rather than the $259,000 to buy it outright. You'd need to drive it on 150 dates before it would make more sense to buy it than rent it. If you don't get a second date with that particular person, you've saved yourself a lot of money. If you do get twenty more dates with the person and then marry him or her, you've still saved $224,000 for your house down payment. If the person objects that you seduced him or her with a false display of wealth—a rented rather than owned Ferrari—you can always respond, "So you married me just for my money?"

You can rent almost anything these days: housing, vehicles, tools, computers, electronics, formal clothing, even designer handbags (see Begborroworsteal.com—"a Netflix for bags"). You can, de facto, rent almost any other durable consumer good by buying it retail and then returning it within thirty days for the full purchase price. (Think of the restocking fee as the rental cost.) Books, music, and movies can be often rented for free from an underused institution called the public library. Indeed, a good rule of thumb before buying anything new is, if possible, rent it for a week, see how you like it, and see whether it has the signaling value you expect.

Some costly items that can't be rented can still be time-shared, such as vacation homes and business jets. If Netjets fractional owner-ship is good enough for Warren Buffett and Bill Gates, it's probably good enough for most of the other 950-odd billionaires in the world, not to mention the 95,000 "ultrahigh net worth individuals" with assets of more than $30 million each.

## Buy It Used

Let someone else absorb the huge depreciation that afflicts almost every new purchase in the first couple of years. Let him buy the new car, and get it at one-third of the price from him after five years, when his compulsive novelty seeking turns to the next bright new thing.

To take advantage of this strategy, one must overcome the irrational premium we put on pristine condition, and the irrational distaste we feel at the pre-owned. For example, one can get almost all clothing from thrift shops. In rich towns and suburbs, thrift shops are full of rarely worn, well-made, premium-brand clothes, including good cotton, linen, or silk shirts by Armani or Claiborne for about $5, which were more than twenty times that price new. (The quality, not the brand, is the point here.) Mostly, the clothes are only there because their original purchasers got fatter faster than expected, so couldn't wear them anymore.

To feel comfortable buying used clothes, you just have to remember two things. First, "new" clothes have not been draped onto the retail store's rack by antiseptic angels who flew straight from an ISO Level 1 clean room. They have probably been handled by dozens of young Asian females in export-processing zone sweatshops, and transported for weeks in dusty, rusty cargo containers, and unpacked by store clerks who may not have washed their hands after their Taco Bell lunch break. The clothes are thus new not in the sense that a fresh-hatched baby newt is new, but in the sense that they have existed for only a few months, traveled only a few thousand miles, been handled by fewer than forty people, and been tried on by fewer than five customers.

Second, you must remember that "used" clothes were simply bought and worn by humans who were genetically, biochemically, and dermatologically just like you, who mostly avoided stains and washed them regularly. The thrift store has already disinfected and cleaned them beyond any requirements of germ-phobic obsessive-compulsives. If you are a misanthrope at heart, the thought of wearing other people's clothing, like the thought of their existing at all, may make your toes curl and your teeth clench. But if you have a reasonable sense of solidarity with your species, wearing clothes pre-owned by strangers will seem no more offensive than wearing clothes pre-owned by siblings or best friends. (Of course, one draws the line at underwear and socks, which aren't sold at most thrift stores.)

These principles—"new" isn't really new, and "used" just means used by people nearly identical to you—hold even more strongly for less-intimate goods. Should you really care that different human hands gripped the used car's steering wheel for fifty thousand miles before yours did? That the Makita sliding compound miter saw already shed some other home carpenter's blood before you bought it from him? That the used bookcase held Tom Clancy novels before your Proust? Irrational disgust, and the fear of symbolic contagion by socially inferior or ethnically different pre-owners, are great enemies of rational consumer frugality.

## Buy It in Generic, Replica, or Trickle-Down Form

The rise of Asian manufacturing, combined with the poor enforcement of trademark law in Asia at present, means that high-quality generic and replica forms of many products are available. As we saw earlier, midprice replica luxury watches work about as well as real luxury watches, and are virtually indistinguishable without a jeweler's loupe and expert knowledge. More generally, technical and fashion innovations characteristic of premium brands are trickling down to mass-market brands with ever-greater speed. This means that premium branding is becoming ever less distinctive as an objective marker of product quality and novelty.

## Make It Yourself

If we bought products purely for their use value, the division-of-labor principle would imply that making things yourself is almost never rational. Someone else could make it far more efficiently than you could, since he already has the workspace, tools, and skills. However, the fact that we often buy products for their display value implies that it is very often sensible to make things yourself, since self-made items can be much more effective trait displays than other-made items. Of course, this is the unconscious function of hobbies and crafts, and explains why rationally specialized professionals, whose time on the open market is worth more than $100 an hour, devote evenings and weekends to making things that a Chinese factory worker could make in about 6.3 minutes for twelve cents. Hobbies and crafts allow one to display intelligence, creativity, and conscientiousness across a broader set of domains than one's formal work permits. They also result in physical objects more easily appreciated by nonspecialists (such as raku serving bowls or macramé swimwear), rather than professional achievements in corporate tax law or amorphous silicon engineering that are beyond the comprehension of friends and neighbors. Economists may feel that house spouses who frequent Hobby Lobby, or celebrities who wash their own cars, are contemptibly ignorant of Adam Smith's great division-of-labor insight. Yet their own spouses, children, and neighbors may appreciate fewer self-righteously rational judgments and more home-baked chocolate-chip cookies.

Making things yourself is easiest, most frugal, and most effective as a display when the materials and tools required are cheap, when the skills are easily learned but seldom mastered, and when the things made are actually useful, beautiful, and visible. You can create presentable dining room tables and bookcases given some hardwood stock, a simple table saw, a basic knowledge of carpentry, and a stock of Band-Aids. Cooking, sewing, jewelry making, and pottery can yield impressive self-made items for similarly low start-up costs. Even when an item cannot be made from scratch, the components are often available for home assembly, as in the case of personal computers. You can buy the case, motherboard, power supply, hard drive, and so on, wire them together, and have a truly personalized product that displays your arcane wisdom about the mysteries of the electron.

Less suitable products for self-manufacture include scuba gear, carabiners, airplanes, pacemakers, and artillery. I also cannot recommend making anything at home that requires thermal lances that burn above 10,000 degrees Fahrenheit, or things that rotate faster than 10,000 rpm. Safety first—unless you feel obligated to display your capacities for risk taking and limb regeneration.

Consumerism's Cthulhu-like tentacles have, of course, reached right into all home-manufacturing activities, and transformed them into runaway consumption domains in their own right. Hobby magazines proclaim that every new pursuit requires purchasing a new set of conspicuously costly, wasteful, precise, reputably branded tools, supplies, and accessories. It is easy to become convinced that, to make a bookcase that would cost $80 from IKEA, you must buy a $1,000 carpenter's workbench, $600 table saw, $500 biscuit joiner, and $300 router (ignoring my advice above about high-rpm tools). Indeed, this is what most guys do, and this is what sustains Home Depot and tool brands such as Craftsman, DeWalt, and Makita. Luckily for you, some of your friends and neighbors have probably already fallen for the delusion that their social status and sexual prospects will be magically enhanced by the existence of high-powered, premium-brand woodworking tools in their garage. You can go borrow them; it will save you money, and reduce their guilt about the tools' never being used.

## Commission It from a Local Artisan

Some things are too hard to make yourself, but can be made by local specialists. This often yields a huge display advantage at only a small cost premium above the mass-produced version of the product. A custom item, designed or specified by oneself, and handmade by a local artisan whom one met face-to-face, is a much more distinctive possession or gift. It displays one's resourcefulness, creativity, taste, and social skills in collaborating with the artist. For example, when buying jewelry for a significant other's birthday, one could always go to the mall at the last minute and grab some ten-karat die-struck abomination of a ring at an 80 percent markup. Or, one could plan ahead, talk to a local artisan, collaboratively design something unique, and have it hand-fabricated in twenty-karat gold. The materials will cost more, but the markup will be lower, and the ring's quality will be higher. Also, the narrative value of giving something unique will be meltingly romantic. Commissioning likewise displays very high creativity, resourcefulness, commitment, and conscientiousness. Indeed, a personally commissioned handmade ring in sterling silver may elicit more gratitude than a mass-produced platinum band.

As another example, many families buy mass-designed houses built in alienating new suburbs by huge developers. The structures are designed to the lowest common denominator of taste in the current fashion, so their aesthetic value depreciates quickly. They are built to poor standards—two-by-four stick lumber and half-inch Sheetrock on concrete slabs—so their physical integrity deteriorates quickly. The houses are not supported by adequate investment in surrounding infrastructure—roads, parks, schools, well-planned retail—so their quality of life depreciates quickly. The result is that in many communities, five-year-old houses have lower equity value than new ones. A good alternative is to commission a distinctive new family house from an up-and-coming local architect on a vacant plot in an established community. The build cost per square foot may be slightly higher than for a mass-designed developer house, but the display value—and home equity—per dollar spent will be much higher. Instead of moving into a house built by nameless, faceless workers, you can move into a house that you codesigned with an architect who might become a friend, and a house that you saw being built by local workers whose names you'll learn and whose workmanship you'll admire. You'll also learn much more about the house, so its features and functions can be more knowledgeably appreciated by you and discussed with others. Whereas others live in houses they understand only superficially, you'll be able to understand all the systems—foundation, framing, roofing, flooring, electrical, plumbing, HVAC, storage, security, decorating—as functional wholes. You'll maintain them better and repair them more easily. And, as the architect's reputation grows, your house's value will increase. This way of living makes a much more effective social display, because it grows social and narrative roots deep into one's local community, and so demonstrates one's creativity, openness, agreeableness, and extraversion much more credibly than buying a prebuilt mass-market house, which requires nothing more than a down payment, a decent credit score, and gullibility.

## Wait Three Years Before Buying New Technologies

Technical innovations are terrific—they drive economic progress, increase quality of life, and give us new things to talk about. But you don't have to be one of the early adopters who pay for the new technology's research and development. Companies usually try to recoup the cost of innovation through very high premium pricing of new technologies. These are bought most avidly by that golden market segment, the eighteen-to-thirty-four-year-old single male demographic, because those are the humans who invest the highest proportion of their money, time, and energy into mating effort, including conspicuous consumption. Then, after the horny rich boys all own the technology, the companies gradually lower their prices and enter the mass market. This is when LCD TVs get 30 percent cheaper every six months. The price lowering must be gradual enough that there is no obvious point where everybody suddenly considers the new technology a bargain—that would not keep consumer demand nice and steady. This gradual price lowering makes it difficult to give any simple rule of thumb for how long to wait before buying a new technology. However, two or three years seems to work for consumer electronics; by that point, prices have usually dropped at least 80 percent since introduction, and reliability problems and format wars (VHS versus Betamax, HD-DVD versus Blu-ray) have usually been resolved. For other types of innovation, the optimal wait time depends on the speed with which the innovation trickles down into mass-market products, which usually depends on how well the patent holder markets and licenses the innovation. If electronic stability control sounds like a good car-safety innovation to you, don't get it as soon as Mercedes-Benz introduces it in their S-class sedans at a $5,000 price premium. Wait until Mercedes licenses it to Subaru and Subaru passes along the true marginal cost of the system per car—about $110—to buyers.

Waiting is hard. For those who enjoy technical innovations, it is useful to remember this: for every bleeding-edge novelty that gets massive press coverage and early-adopter attention now, there are several technologies that were bleeding-edge five years ago that are just now getting broadly licensed and showing up in mass-market products. Those are the novelties to seek—the underadvertised bits of progress that pervade every domain of product land. They may not be objectively new in world-historical terms, but if they are new to you and the people to whom you are targeting your trait displays, that is new enough. For example, one might go to the Consumer Electronics Show and feel impressed by the new ZPrinter450, the world's first "affordable" color 3-D printer, which can create complex 3-D plastic objects from computer specifications for just $40,000. But before getting that second mortgage, consider the computer-controlled electromechanical innovation that was cutting-edge just a few years ago: the Janome Memory Craft 9500 embroidery machine, which can stitch almost any pattern you design on the computer, in any colors, on any material. Its street price has recently dropped another 30 percent, to about $1,500. The ZPrinter450 is cool, but so is the Memory Craft 9500. In fifteen years, nobody will remember which was invented first. And probably no one in your set of potential mates or friends has heard of either one, so they have equal novelty value as trait display signals.

## Ask to Get It as a Gift

If you decide you simply must have the product new, branded, at full retail price, put it on your birthday or holiday wish list, and let somebody else buy it for you. Your friends and family never know what to get you anyway, so you might as well let them know, and enjoy the anticipation of waiting for it. Gift giving is central to human social life; it reinforces bonds of affection and reciprocity. Receiving something as a gift from someone worth talking about, under memorable circumstances, hugely amplifies its personal display value by testifying to one's lovability and popularity.

## Acknowledge the Display Premium Built into Most Retail Products

If you still must have the new, branded, full-retail-price product, and nobody will give it to you, it is at least wise to calculate and acknowledge whatever display premium is built into the product's retail cost. For example, earlier in this book I discussed how most basic electromechanical products, ranging from small appliances and electronics through large appliances and cars, cost about $10 per pound. If an electromechanical product costs more than about $12 per pound, it is probably a premium brand or newly marketed item that includes a substantial conspicuous-consumption surcharge above its manufacturing cost. For example, a Toyota Camry costs about $7.00 per pound ($24,000 for thirty-four hundred pounds), whereas a Lexus LS costs about $16.50 per pound ($71,000 for forty-three hundred pounds). The Lexus luxury-car premium is obvious from its much higher cost per pound, despite its being assembled of similar materials to a similar level of precision by the same parent company. So, if you must have the Lexus, that's OK, as long as you consciously accept two things: (1) apart from its higher mass, you are paying an extra $40,000 for the Lexus badge, and (2) everyone who sees you driving the Lexus, and who has read this book, will assume that you could think of nothing in the world more creative, kind, or conscientious to do with $40,000. To the 2.8 billion people who live on less than $2 a day—or anyone who cares about them—this lack of imagination might seem a bit blinkered.

## Taming the Centrifugal Soul

Some common themes emerge from these slightly whimsical suggestions. One is that buying new, real, branded, premium products at full
price from chain-store retailers is the last refuge of the unimagina-
tive consumer, and it should be your last option. It offers low narrative
value—no stories to tell about interesting people, places, and events
associated with the product's design, provenance, acquisition, or use.
It reveals nothing about you except your spending capacity and your
gullibility, conformism, and unconsciousness as a consumer. It grows
no physical, social, or cultural roots into your local environment. It
does not promote trust, reciprocity, or social capital. It does not expand
your circle of friends and acquaintances. It does not lead you to learn
more about the invention, manufacture, operation, or maintenance of
the things around you.

Retail spending reveals such a narrow range of traits: the capacities
to earn, steal, marry, or inherit wealth, and the perceptual memory
and media access required to spend the wealth on whatever is advertised most avidly now. The alternatives listed above try to minimize
retail spending not just to save money, but to maximize trait display
power. They do not renounce the emotional centrality of narcissistic
self-display in human life, but they redirect it into tactics that are both
cheaper and more effective.

Because many of these tactics require less money but more time,
they might seem inefficient to economists who are always weighing
one's opportunity costs: in principle, every hour of leisure time taken
costs one an hour of work uncompensated. If a male lawyer who can
bill $300 per hour can buy a new shirt in ten minutes from Neiman
Marcus for $100, rather than searching through a thrift store for forty
minutes to buy a used shirt for $5, he should logically buy the new
shirt, which saves $150 in potential billable hours. This is true as far
as it goes, but it ignores the larger context of the retail purchase. A
fair comparison must take into account the total shirt-research and

shirt-purchase time. If the lawyer spent fifty minutes browsing *GQ* magazine to see which leisure shirts are in fashion, twenty minutes choosing which one he prefers, ten minutes calling local stores to find out who has it in stock, sixty minutes driving to and from Neiman Marcus, and forty post-purchase minutes defending his *GQ* reading and shirt purchasing against his wife's aesthetic skepticism, then he's really spent three hours on the shirt purchase, and its cost is really $1,000: $100 retail cost and $900 opportunity cost. By contrast, he could have driven to the nearest thrift store, used its logical arrangement of stock by garment type, size, and color to quickly identify some interesting shirts, tried them on, picked one, and bought it, in a total shirt-purchase time of about one hour. If his wife doesn't like the shirt, no problem: it only cost $5. It could be burned impulsively on the barbecue to display his respect for the wife's superior aesthetic judgment, and she would love him for it, and they could have connubial canoodling for two whole hours, and he would still come out ahead. Plus, the whole episode would make a great dinner-party story. If such purchases are seen clearly in their entire economic, temporal, social, and trait-display contexts, then the more time-demanding tactics listed above will often prove not just more romantic, but more rational.

There is a deeper problem with habits of retail spending that try to minimize time—they fail as a costly, reliable signal of one's dedication to a particular person (in the case of gifts), or to a particular acquisition (in the case of things bought for self-display). For the educated professionals most likely to be reading this book, time is often more precious than money. Our checking account might be comfortably bloated, but we don't have any time after the demands of work, family, household, and sleep are met. Or, we have a little leisure time, but no intelligence or energy to do anything more complex than watch *Desperate Housewives*. Under these time-limited conditions, giving gifts and acquiring goods that require high personal time investments are much more credible, impressive signals of generosity and taste. They indicate that we gave up something significant (some work time, kid time, or Teri Hatcher–ogling), because we cared more about someone, something, or some trait display.

This is where making something yourself speaks much more loudly than a premade thing's retail price, and a little perspective taking shows why. Consider which kind of Valentine's Day gift you would rather receive from someone whose affection you crave but doubt: (1) a dozen red roses ordered online, and delivered with a generic "Loads of love" computer-printed note, or (2) a customized sonnet, composed by the loved one, written on handcrafted paper, and delivered with home-made chocolate truffles decorated with the candied violets that you once mentioned liking in an e-mail eight months earlier? The latter would constitute a hugely more compelling display of personal interest (affection, thoughtfulness, lustful intent) and of personal traits (cre-ativity and conscientiousness in the domains of poetry, calligraphy, cuisine, and pulp-and-deckle work). You can tell that the latter is more compelling because it also makes a better story to tell your family and friends.

In fact, the whole valence, significance, and trait-signaling power of any given product can be radically altered by the stories we tell about it. Suppose you're a man who meets an intriguing woman in the Café des Hauteurs atop the Musée d'Orsay in Paris. You chat about the nearby Renoirs for a few minutes, and then notice she's wearing a fine gold filigree ring with a star sapphire. You praise its workman-ship and inquire about its origin, and she replies with one of the following answers (which allow the following trait inferences, in parentheses):

- "I got it from a Zales mall store in Albuquerque" (normal American tourist)
- "I got it from a Zales mall store in glamorous Albuquerque, as a treat for finishing Habermas's *Theorie des kommunikativen Handelns*" (intelligent, open, intellectually pretentious but geographically self-deprecating American tourist)

- "I inherited it from my grandmother Valya, who commanded a T-34 in the Great Patriotic War" (nostalgic, family-oriented Russian tourist familiar with World War II tanks)
- "I got it at the Glastonbury music festival, when Moloko played there back in 2000, to celebrate my one-hundredth Ecstasy trip" (open, impulsive, agreeable counterculture English raver)
- "I made it myself. Working with the molten gold and hot wax helped distract me from my divorce four months ago" (skilled, resourceful, thermophilic survivor, newly single)
- "It's a tracking device that the aliens put on when they abducted me. Sadly, if I tried to take it off to show it to you, the gram of antimatter inside would vaporize everything within eight kilometers" (delusional schizophrenic with semiaccurate understanding of $E = mc^2$)

In each case, the conversation sparked by the ring is hugely more informative about the wearer than the product's physical features or apparent cost. Without human language to weave stories and social connections around products, the products are as mute and enigmatic as Duchamp's readymade urinal. With language, the products take their rightful place as conversation starters, ways of breaking through modern alienation and wariness of strangers.

Good talkers don't need such products to meet, dazzle, and endear people. The great conversationalists—Samuel Johnson, Madame de Staël, Denis Diderot, Virginia Woolf, Isaiah Berlin, Gore Vidal, Christopher Hitchens—have always been able to display their intelligence, personalities, virtues, and knowledge through language alone. Rather than buying more products, almost everyone can improve his conversational skills more effectively through Dale Carnegie's "six ways to make people like you" from *How to Win Friends and Influence People*: remember the person's name, listen well, smile, make the other person feel important, talk in terms of the other person's interests, and become genuinely interested in the other person. Few people master these skills, so many people strike others as self-centered and boorish.

Most of us need to listen less to our iPods and talk more with those nearby.

## The Promise of Mass Customization

We yearn to display our distinctive traits—our individuality, uniqueness, identity—in as many of our possessions as possible. Those who can afford customized products tend to buy them: whether in the form of tailored clothes, commissioned portraits, or custom yachts, the wealthy prefer the bespoke. However, many products would simply be too complicated and expensive to commission from artisans. An architect can custom build a house, and a jeweler can custom cast a ring, but what about higher-tech goods, such as cars and iPods? Fortunately, some new technologies of mass customization will make it ever easier for a wider range of customers to "express their individuality" (display their traits and preferences) through products made uniquely for them.

The idea of mass customization is that companies use the efficiency and precision of mass production technologies to make unique products for particular customers, according to their specifications. The "batch size" for any given production run is reduced from thousands of units to a single unit. This build-to-order system requires shifting to a leaner, more flexible production line, and demands real-time control of the supply chain. However, it benefits producers by reducing wasteful inventory, and it benefits customers by delivering a product that fits their preferences more precisely.

For example, if Rolex were set up for mass customization, and you wanted a picture of your newborn daughter etched onto the face of a new Rolex President watch, you could visit their website, upload a digital photo of your daughter, select from a few options, pay for the watch, and get it in the mail within a week. Or, if you wanted the paint on your new Harley-Davidson Fat Boy to match the color of green M&M's, you could get the Pantone color reference number of green M&M's from Mars Incorporated, and e-mail it to the Harley-

Davidson factory. The companies might charge a high premium for such customization, but the customer would get an individualized product worth talking about.

For mass customization to work, the production technologies involved must be very flexibly controlled by computer systems that are open to direct input from customers. The process is easiest for products that can be surface detailed using digital printing or etching technology (watch faces), mixed from standard materials (paint pigments), assembled out of standard modules (Dell computers, Renault cars, or package holidays), or cut from thin materials using computer-controlled lasers or routers. Typical applications are in product categories such as customized books, posters, or T-shirts, mixing custom-perfumed toiletries, or weaving custom textiles (as has been done with programmable Jacquard looms since 1801). The site Colorwarepc.com is already offering individually color-customized smart phones, media players, and computers. We are not far from bookstore customers' being able to order a special edition of *The Selfish Gene* printed on cream-colored vellum pages, bound in turquoise leather, to be picked up within a few hours. The next phase of mass customization is likely to include custom assembly of custom-produced parts, as would be required to make a serviceable replica of one's favorite Danish-modern oak chair that has now broken, or of Andúril, Aragorn's sword from *The Lord of the Rings*, which was broken but is now reforged.

It will be much harder, but more rewarding to consumers and profitable to businesses, to harness the flexibility of potent new methods in computer-controlled manufacturing, robotic assembly, rapid prototyping, 3-D printing, and automated product testing. These are the technologies that might allow a customer to buy an iPod shaped like their favorite species of tropical fish, or a BMW M5 that looks like the Batmobile. Such applications raise new issues of product safety and reliability—an iPod shaped like a lionfish might cause piercing or stabbing injuries; fins might break off and be swallowed by toddlers or Shih Tzus; lionfish-phobics might sue for emotional distress. Nonetheless, mass customization will vastly expand the customer's freedom to create a unique product that signals distinctive traits and tastes at a reasonable price. Nobody yet knows, however, how marketing and branding will work for such mass-customized products, or how they will ultimately revolutionize consumer behavior, material culture, and trait display.

# 16

# The Will to Display

SUPPOSE I HAVE been right about consumer narcissism—its evolutionary-psychological roots, modern manifestations, and pervasive failures to deliver the trait-display benefits we want? What can we do about those failures, not just at the personal level as addressed in the previous chapter, but at the social, political, and cultural levels?

Friedrich Nietzsche argued that most human action is driven by a "will to power"—we try to thrive as organic beings through dominance, daring, and resource appropriation. His was a vision of lone alpha males gaining existential insights on Alpine peaks, then descending to enjoy the fruits of public acclaim: book deals, harems, and *weissbier*. In *Spent* I have argued that among highly social primates like us, who reliably rebel against Hefeweizen-swilling philosopher-kings, the evolved "will to display" (gaining fitness benefits through prestige and status) can be even more important than the "will to power" (gaining fitness benefits through dominance and threats). Consequently, a central question for social policy is this: How can we harness the will to display so it makes us all happier? Strategies to change our runaway consumerist culture can work on many levels, but the least effective levels are often overemphasized in social-policy debates and political activism, while the most effective levels are often overlooked.

## Asymmetric Warfare

Opting out of mainstream consumerism is like entering a new world of asymmetric warfare. In traditional symmetric warfare both sides play by certain tacit rules of engagement. You line up your phalanxes, musketmen, or tanks, and we line up ours, and both sides fight it out until one concedes or flees, and the other declares victory. In asymmetric warfare, the side that is weaker by traditional criteria seeks victory by using new tactics or technology. The British longbowmen defeated the French knights at the Battle of Agincourt in 1415 by firing volleys of arrows from absurd distances, rather than waiting honorably to be squashed by the cavalry charge. The American revolutionaries of the 1770s defeated the British redcoats through guerrilla tactics—cutting supply lines, harassing troops, sniping from a distance, and simply refusing to line up accommodatingly to get shot by the larger British forces. The Vietcong defeated the Americans in the 1960s through similar tactics. Al-Qaeda terrorists on 9/11 infuriated the Pentagon by hijacking our airplanes, rather than buying their own from our arms dealers (as Saudi citizens are supposed to). Innovations in asymmetric warfare are always initially considered to be treachery and terrorism by the side that believes it is stronger according to traditional criteria. In retrospect, such tactics are inevitably reframed as natural historical progress in the efficient conduct of warfare.

Likewise, every signaling innovation in human culture is at first considered unfair and disreputable, at least by those who excelled at the previous signaling game. Medieval lords were no doubt driven nuts by the minstrels and troubadours who used musical innovations (iso-rhythmic motets, polyphony, even madrigals!) to seduce their wives and daughters, rather than winning them by the traditional methods (physical force, economic oppression, religious indoctrination). Elvis wasn't playing fair by wiggling his hips and sneering, and Miles Davis wasn't playing fair by being so damned cool, handsome, and talented. From the viewpoint of social competitors and sexual rivals who "play fair" by getting formal educations, working full-time jobs, and paying full retail prices, any of these alternative ways of displaying one's personal traits seem like cheating. However, from the viewpoint of rational individuals seeking maximum social and sexual status at minimal cost, all these tactics were wonderfully liberating. Indeed, such signaling innovations seem to drive most of the progress in the technologies, ideas, and institutions that we call civilization.

## Why Not Trait Tattoos?

Premise 1: Conspicuous consumption is a wasteful and ineffective way to display our psychological traits to others. Premise 2: Those traits can be assessed fairly accurately from a few minutes of informal social interaction, but can be assessed even more accurately through formal intelligence and personality tests. These premises suggest an obvious solution, as mentioned in a previous chapter, to the problem of runaway consumerism: encourage everyone to get his Central Six traits evaluated using the best available tests from reputable testing institutions, which could then tattoo the validated trait scores onto the entire population's foreheads. That way, everybody could see at a glance who they're dealing with and how they're likely to behave. This signaling system would obviate the need to display the Central Six through conspicuous consumption.

To people who consider conspicuous consumption more dignified than facial tattoos, such a trait-tattoo system may sound like an outrageous affront to human dignity. To people like me who live in Albuquerque, where many people are poor and many do have facial tattoos, it doesn't seem so absurd. The question is whether it would work, both technically and socially.

We've already discussed the problem that trait tattoos would not be reliable if individuals themselves were responsible for telling the tattoo artists what their Central Six scores were. However, it would be fairly easy to make the trait tattoos technically reliable and hard to fake, given some modest improvements in trait-testing institutions and technologies. The trait tattoos would need to be inscribed by a reputable testing agency, in some sort of conspicuously precise form that would be no easier to counterfeit than currency. The testing institutions would need to be honest, transparent, closely monitored, publicly accountable, and hard to bribe. The tests would need to be objective, reliable, valid, and in the public domain, subject to continuous improvement through open scientific research and peer review. The tests would need to be administered to each individual from adolescence onward at regular intervals, depending on the empirically observed stability of trait scores across the life span. (Studies show that the Big Five become progressively better differentiated, more coherent, and more stable from ages ten to twenty, for example.) If someone doesn't want his trait scores tattooed on his forehead, he would ask for the scores to be held in private as his personal property, to be revealed only at his request to a potential employer, neighbor, mate, or any other person, who could be sent an unfakeable, electronically authenticated test-score report.

Intelligence researchers already know how to measure general intelligence accurately, and the Educational Testing Service already knows how to make IQ-type test scores hard to fake. It would be more challenging to measure the Big Five personality traits through new objective tests, rather than via the self-report questionnaires currently used. At the moment, the closest things we have to such objective personality measures are certain documents requested by employers: the criminal background check and college transcript as fairly good indicators of conscientiousness; the recommendation letters, personal references, and job interviews as fairly poor indicators of agreeableness and stability. We could, in fact, do much better, using current test theory and methods, plus a little imagination. For example, an individual's Big Five traits could be assessed quite accurately by averaging anonymized peer ratings from a large sample of that individual's neighbors, acquaintances, co-workers, old friends, and ex-lovers. These people have observed the individual across a wide variety of situations, states, and moods, and they have unconsciously learned to see through the individual's ways of presenting a good facade in short-term interactions. By aggregating their knowledge, the information-sharing power of gossip and reputation in small-scale human societies could be replicated at the scale of modern economies. Everyone else in the world could know as much about each individual as that individual's whole social network already knows.

Reliable personality trait tattoos would make life easier for the virtuous and harder for the vicious. At the moment, the meek and humble often lose out to the assertive and histrionic in competitive employment markets, mating markets, and social clique-formation. Wallflowers would be favored more often if their agreeableness and stability were more visible, and as people learned how useful such traits really are. Conversely, life would be harder for psychopaths, who can be highly charming in the short term, but who make a lot of long-term enemies. Their disagreeableness and unconscientiousness would become all too apparent, and they would be shunned, at least by sensible folks.

Other approaches to objective personality testing could be based on electronic records, brain imaging, or DNA testing. For example, a person's conscientiousness, agreeableness, and extraversion might be revealed fairly clearly by data from his e-mail accounts, mobile-phone accounts, and social-network sites—such as the number of non-spam e-mails, phone calls, and social-network messages received and responded to per month. As researchers discover which brain-response patterns are most closely associated with each personality trait, it should become possible to assess the Big Five by exposing people to various virtual-reality situations during brain-scanning sessions.

A final option would be to go straight to the genotype. Companies such as 23andMe and Cambridge Genomics already offer "consumer genomics" and "personal genetic testing." Geneticists are poised to sequence a person's entire genome for under $1,000 within a few years. I've argued in some recent papers that specific alleles (forms of genes) that are associated with personality traits should be much easier to find than specific alleles associated with general intelligence or general health. Some personality-related genes have already been identified; others are being discovered at a rapid rate. It will be quite a few years before brain imaging or genotyping achieves the personality-assessment accuracy that we can now attain using large samples of peer ratings or actual behaviors, yet eventually, using all these methods together should yield good objective measures of the Big Five.

While trait tattoos would be technically feasible as reliable signals of the Central Six traits, would they be socially accepted by the signalers (tattoo displayers) and receivers (tattoo viewers)? If social norms favored having the trait tattoos, everyone would have social incentives to get them, even if they scored pretty badly on some traits. This is one of the most surprising predictions from costly signaling theory. You might think that only people with very high intelligence would volunteer to get their IQ scores tattooed on their foreheads, and indeed that might be true at first: only IQ-150 geniuses would bother. But then, all the people with an IQ between 125 and 150 would want to show that, even if they're not geniuses, they're a lot brighter than most of the other non-tattooed folks (whose average might now be about 95), so they would follow suit. Then all the people with an IQ between 100 and 125 would want to show that they're at least average compared with all the remaining untattooed folks (whose average is now about 90), so they'd get the tattoos. And so on down the whole IQ range; even the mildy retarded (IQ range 50–70) or their guardians might want to show that at least they're not retarded to a moderate (IQ 40–50), severe (IQ 20–40), or profound (IQ below 20) level. The same reasoning applies even more powerfully to the Big Five personality traits, which don't have such a clearly value-laden dimension. If people are happy to advertise their levels of openness or extraversion through bumper stickers, they might be equally willing to do so through tattoos.

However, it might ultimately be hard for people to accept the trait tattoos as worthy of attention. The problem here is the way that our perceptual systems are wired up to our person-perception systems. We have evolved exquisite abilities to make unconscious inferences about other people's psychological traits, by observing their behavior. But those abilities were adapted to prehistoric conditions, and they "expect" the relevant social information to come in certain natural "information formats" through certain perceptual channels. For example, we can easily judge other people's intelligence if we hear them say very funny and insightful things, or see them produce a very creative and beautiful artwork. We can tell that these perceivable cues of intelligence fit naturally into our person-perception systems because they provoke not just cool respect for that person's intelligence, but hot emotional responses such as awe, admiration, deference, envy, friend-liness, or lust. By contrast, there are hundreds of other perceivable cues of intelligence that may be equally reliable and valid, but that reach our ears and eyes in evolutionarily novel ways that our person-perception systems did not evolve to process so easily. For instance, a complete 3-D MRI scan of someone's brain might convey as much objective information about his intelligence as hearing him play an awesome drum solo, but inspecting a brain scan just cannot inspire as much social respect or sexual attraction as moving one's body to a compelling rhythm. Likewise, a trait tattoo might convey more-reliable information about intelligence than a ten-minute conversation, but it cannot spark the emotions that drive social interaction. There seems to be no easy shortcut through our person-perception systems. We have to feed them the kinds of social stimuli that they evolved to expect, and institutionally validated trait tattoos are not among those stimuli. By contrast, our person-perception systems seem surprisingly happy, after just a couple of decades of consumerist socialization, to process information about an individual's conspicuously wasteful, precise, and/or reputable possessions, and to make personality inferences on their basis.

Although institutionalized objective tests of people's Central Six traits may never replace more natural forms of information about peo-ple, they may still prove useful in some situations. For example, how much would you pay before a first date to get accurate, objective, elec-tronically validated information about a suitor's true personality traits, based on extensive peer ratings, computer and phone records, brain scans, and genotyping? If I were single, I would buy that for a dollar. If I were a woman who had suffered through too many dates with psychopaths, narcissists, depressives, egoists, and date rapists, I might pay a lot more. When my daughter starts dating, or thinks she wants to marry someone, I will sorely lament the lack of such a service. Also, I would love for doctors, lawyers, architects, car mechanics, house-cleaners, and real estate agents to post their validated IQ and Big Five scores in their Yellow Page ads and LinkedIn Web pages. It would save everyone a whole lot of time, trouble, and money. We wouldn't need to ask for references from three previous clients if 150 of their friends, relatives, in-laws, neighbors, and ex-spouses have already rated their conscientiousness, agreeableness, and stability. The same sort of consumer feedback that has revolutionized shopping for goods on the Internet (based on percent-positive feedback from customers) could revolutionize shopping for personal and professional services.

One might think that common sense alone would suffice to help people avoid narcissistic dates or disagreeable doctors. But if such individuals were that easy to detect and avoid, those heritable traits would have already been eliminated from the gene pool. The endless co-evolution between our truth-seeking person-perception abilities and our deceptive trait-display tactics never reaches a point where everyone's first impressions are always accurate. We could learn to use objective test scores to supplement our natural but fallible first impressions, just as we use calendars when planning or speedometers when driving. The result would be mass social transparency at the level of the Central Six traits. At least at a conscious, rational, practical level, everyone could know almost everything worth knowing about everybody else. They might still feel subconsciously attracted to bad boys with psychopathy or goth girls with borderline personality disorder, but they would at least have a chance of sensibly overriding that attraction. Mass social transparency sounds frightening and embarrassing, but it is what humans have been striving for ever since the prehistoric development of gossip, reputation, "face," and status symbols. It would allow at least some rational people, some of the time, to choose their friends, mates, co-workers, and neighbors more quickly and accurately.

## Prerequisites for Buying Certain Products

Trait tattoos might not work socially because they don't express traits in the right information formats to be processed easily by our person-perception systems. However, there may be more user-friendly or easy-to-learn ways that individuals could display their validated trait scores.

For example, companies could sell certain products only to consumers who have a certain minimum or maximum score on one or more of the certain Central Six traits. Hummer dealers could advertise that the "Party Animal Red Pearl" paint color is available only to customers who score in the top 5 percent for extraversion. Customers who want to display their unusually high extraversion through that bright red color would have to electronically validate their extraversion score at the dealership before they could sign the purchase agreement. In this way, Hummer could guarantee that Party Animal Red Pearl becomes a reliable signal of friendliness, self-confidence, and ambition. Or, Lexus could sell the "Mensa Quartz Metallic" color of the LS 460 only to customers whose validated intelligence scores are high enough for them to join Mensa International (IQ 130+, or the top one in fifty). The more exclusive "Prometheus Glacier Pearl" color could indicate an IQ above 160 (the top one in thirty thousand) –the qualification for joining the Prometheus Society.

Product prerequisites could include not just the Central Six traits, but any information that customers would be willing to divulge by giving the retailer access to electronic records. This information could include age, sex, residence, education, employment, financial records, marriage records, medical records, paternity tests, church attendance, political-party registration, or previous purchase history—any data that might be relevant, and that could be somehow validated. For instance, Durex could sell a "Clean Submarine" brand of motorcycle helmet only to males who have regularly tested negative for all sexually transmitted diseases. Or, the Mormon Church could sell a "10 for 10" model of laptop computer only to members who have paid their 10 percent tithe on gross income regularly for at least 10 years.

Products could even require some socially or sexually desirable combination of traits. The "Mr. Right" model of Sketchers casual shoes could be sold only to single males age twenty-five to forty who have high intelligence, high conscientiousness, and high stability—

the "good dad" traits that women seek for marriage. The "Happy Fluffer" jacket from Spyder Skiwear could be sold only to single females age eighteen to twenty-nine who have high openness, high agreeableness, and high impulsivity (low conscientiousness)—the cues of sexual availability that men seek for short-term mating.

How exactly would this system work? Trait prerequisites and a product barcode number would be registered for each of these products with an International Test Score Database, which would also contain test scores for every customer who wants to participate in the system. The test scores would be based on the same sort of objective measures mentioned for trait tattoos—intelligence tests, peer ratings, electronic communication records (for example, number of Facebook friends), perhaps even brain scans and genotypes. At the point of purchase, the customer would present his International Test Score ID card, and the cashier would make sure the ID card photo and biometrics matched the customer. The cashier would scan the product's barcode into the point-of-sale computer connected to the database, which would authorize the sale if that customer's trait scores matched those required for the purchase. The database would hire legions of "mystery shoppers" to make sure cashiers were not making unauthorized sales that violated the product prerequisites and thereby undermined the product's signaling power.

At first glance, the "product prerequisite" system might seem counterproductive. Why would a company want to restrict its potential target market by selling only to customers who have some set of specified trait scores or other background information? Think about it for a minute. This would be by far the easiest way for a new product with a particular brand personality to position itself in relation to customers with certain personality traits and/or market segments with certain demographic traits. The signaling power of the product—and its desirability to consumers, and profitability to marketers—would be vastly increased by such requirements.

The product-prerequisite idea has some clear implications for marketing new services, such as private spaceflight. The astronauts of the

1960s were deeply respected not just because they had orbited the earth a few times, but because they had undergone the most rigorous selection regimen ever imposed on job applicants. If Virgin Galactic makes private spaceflight too easy, passive, and comfortable, so that anybody with the cash can fly, its customers will lose that astronaut mystique, and Virgin will attract fewer customers to its spaceport-under-construction near Truth or Consequences, New Mexico. Far better to emulate the NASA selection system in the private sector, and allow only very healthy, intelligent, and stable candidates to fly Virgin Galactic. Let a lower-status brand, perhaps Carnival SkyCruise or RyanRocket, cater to the octogenarian-billionaire and chubby-heiress set, who want to brag about surviving a low-g liftoff and a few safe low-earth orbits. The space-travel market will be glutted by early entrants such as Planetspace, Rocketplane Kistler, Armadillo Aerospace, Blue Origin, XCor, and Bigelow Aerospace, but after the inevitable shake-out of bankruptcies and mergers, expect only the most selective brands to survive.

## A Government War on Bling?

If trait tattoos aren't socially feasible and marketers don't yet appreciate the benefits of product prerequisites, how else could people display their Central Six traits without conspicuous consumption? Social scientists often reach straight for the blunderbuss of government policy whenever they see a social problem that needs fixing. The recommended solutions usually create new programs to minimize the problem's human costs, and new laws to criminalize the alleged roots of the problem. This usually leads to new bureaucracies with vested interests in perpetuating the problem so they can continue to ameliorate its costs and punish its perpetrators. Moreover, individuals and nongovernmental institutions, who learn and adapt far faster than governments can, always end up exploiting the hidden loopholes and incentives structures of the new regime. The transient 1950s Communist threat yielded the permanent Pentagon bureaucracy, and its exploitation by arms manufacturers. The transient 1980s crack-gang wars yielded the permanent "War on Drugs" and its exploitation by purveyors of "Just Say No" propaganda, home drug test kits, and rehab centers. The transient 9/11 Al-Qaeda attack yielded the permanent "War on Terror" and occupation of Iraq, and their exploitation by Halliburton.

The examples could be multiplied ad nausea. Yet, every time a politician or policy wonk develops a new carrot-and-stick program to solve a social problem, and manages to get it implemented in the managerially flawed, politically compromised, underfunded way that inevitably occurs in real life, he is always surprised by the harmful side effects of his good intentions. The brilliant new policy's failures are then attributed to the flaws, compromises, and underfunding, rather than to the dubious conceit that complex social problems can be solved by simple government interventions.

So, I will not recommend new laws to criminalize consumer narcissism, luxury goods, or status symbols as part of some new government War on Bling. History shows that all such attempts to equalize human status by government fiat are doomed to fail. Make people wear Mao suits, and they will compete for status by waving around a newer edition of Mao's Little Red Book, or beating their teachers with larger sticks during the Cultural Revolution. The instincts for status seeking may have deep evolutionary roots, but the cultural modes of status seeking are far too protean for any government to track. There are much easier and more flexible ways to change human behavior through shifts in the informal, grassroots social norms governing trait display.

## A Little Something Called Civil Society

If government can't humanize consumerism into a more efficient and agreeable system of trait display, what leverage do we have? We Americans may assume that the only alternative to a federal War on Bling is a modest shift in personal lifestyle. As a nation of aspiring plaintiffs, we assume that if it's not in the law, it must reside in the private sphere, where personal tastes are beyond public discussion. This is because we have been brainwashed from birth to ignore all forms of social organization and cultural power that exist at levels of description other than the idealized Constitution, the amoral corporation, and the atomized individual.

Educated people elsewhere recognize that well-functioning societies include a few other factors that can shape human behavior rather powerfully: cultural traditions, social norms, customs, habits, languages, memes, etiquette rules, belief systems, and group identities. These systems of behavioral norms constitute a little something called civil society. They are what sociology and cultural anthropology study, albeit using the most obscure vocabulary possible ("ideologies," "discourses," "hegemonies," "lifeways"). They are what make life in north London different from life in North Dakota. They are what make Amsterdam cooler and happier than Kraków or Karachi.

In the late 1980s through the late 1990s, development economists often touted the "Washington consensus," arguing that poor countries could grow rich simply by setting up free markets, free trade, and stable currencies. After the stagnation of ex-Communist economies, the Asian currency crisis of 1997–98, and the rise of ethnic conflict and religious extremism, it became clear that prosperity requires more than just free markets. First, it requires the rule of law: good governance to enforce fair, stable laws regarding property rights, human rights, and social stability, as evaluated by the World Bank's Governance Indicators. Second, it requires sociocultural traditions of accountability, transparency, morality, and trust in politics and business. Third, it requires behavioral norms of valuing education, ambition, initiative, hard work, politeness, peacefulness, and social networking—in other words, norms that tend to maximize the Central Six traits of workers and consumers.

These social institutions and behavioral norms provide the "operating systems" and "applications" with which all human institutions run—not just governments and corporations, but marriages, friendships, families, neighborhoods, public spaces, cities, professions, careers, games, leisure activities, churches, clubs, charities, and chat rooms. This has always been true for human groups, especially during the past few thousand years in every stratified society with a complex division of labor. Civil society is where the action is, where reasoned argument and new knowledge can most effectively change people's lives for the better.

To effect change through civil society, one must understand and accept the informal systems of person perception, praise, and punishment on which such a system relies. People indoctrinated in hedonistic individualism, religious fundamentalism, or patriarchal nationalism—that is, 99 percent of humanity—are not accustomed to thinking imaginatively about how to change society through changing its behavioral norms and institutional habits. Indeed, the more-extreme advocates of hedonistic individualism, such as many classical liberals, libertarians, rational-agent economists, and counterculture anarchists, tend to reject civil-society norms as a basis for changing or sustaining anything at the social level.

A hidden reason for their skepticism is that civil-society norms must rely on fallible personal judgment and a philosophically incoherent notion of free will. Informal social norms only work if individuals make inferences about the personalities, capabilities, and moral virtues of others by observing their behavior. Such inferences are always based on incomplete information, probabilistic cues, and past experiences, so they are always fallible—and open to charges of prejudice, bias, and stereotyping. Likewise, informal social norms work only if individuals are willing to praise or punish others for observed behaviors and inferred traits that must logically be a joint product of their genes, environments, and accidents. We must be willing to act as if people are admirable for personal virtues and culpable for personal failings—as if free will existed, even though we know that, metaphysically, it does not. (Does this paradox identify an awkward hypocrisy at the heart of human social life? Absolutely. Is there any way around it? Not that I can see.)

These prerequisites for civil-society norms—fallible person perception and a faith in free will—have led to some notorious madness-of-the-crowd excesses throughout history: Puritan witch hunts, Paris guillotines, Klan lynchings, Hutu versus Tutsi machete killings. True, such episodes pale beside the mass carnage inflicted by totalitarian governments (Hitler, Stalin, Mao, Mugabe), but they are deplorable nonetheless. Paralyzed by the dangers of fallible judgment and the paradox of free will, many liberals and academics have explicitly rejected informal social norms as a basis for civil society. We espouse the ideologies of tolerance and diversity, which boil down to an unwillingness to praise or blame anyone for any behaviors. The result is that we have no leverage for effecting social change, except through government intervention.

## The Power of Informal Social Norms

Recent research in game theory and experimental economics has shown that informal social norms can powerfully influence human behavior and sustain human cooperation. This is especially true for systems of socially distributed punishment, in which many individuals impose sanctions on the few who do not behave properly. When individuals live in true societies, with repeated interactions among locally known neighbors who have the power to reward and punish one another, these sanctions work as very efficient, credible deterrents against antisocial behavior. Punishing bad acts is much easier than rewarding good acts, because there are many low-cost ways to impose fitness costs on people (taking away their resources, status, freedom, or bodily organs), but only a few, high-cost ways to give them true fitness benefits (awarding them longer life, extra sexual partners, and babies).

Social norms sustained by the threat of informal neighborly punishment can solve the problems of cooperation in many "games" studied by game theorists, including the iterated prisoner's dilemma and the common-pool resource dilemma. These games will sound tiresomely familiar to economists and utterly obscure to everyone else. But they are important, because they reflect the key challenges of human social life in explicit, analyzable ways. If social norms backed up by informal punishment can promote cooperation in the iterated prisoner's dilemma, they can also, at least in principle, promote peace and happiness in your life.

This has been a very cautious, roundabout way of saying that it's OK to treat your neighbor as a villain if he acts like a villain. In fact, it's your civic duty—in the strict sense that civil society could not function without such informal social punishments and rewards. Rural villages needed busybodies. Modern cities need morally assertive citizens. Formal law, police, courts, and jails have never sufficed to sustain a collective quality of life worth living. Indeed, you don't need written constitutions, corporate mission statements, or personal catechisms if you have a genuine culture—a set of informal behavioral norms—that is tacitly understood and enforced by most people in a society. The British understand this perfectly well, and find it amusing that Americans whimper with existential vertigo if their national, corporate, or individual values aren't written down somewhere on paper. Informal norms must do 99 percent of the daily work of shaping human behavior in socially desirable directions. This principle has been clearly understood by every sane adult in every functioning society for thousands of years; Euro-American liberal academic subcultures of the late twentieth century are the singular exception. Until the power of informal social norms is more broadly and consciously appreciated, we'll continue to overlook the single most potent way that we can change society in general, and consumerism in particular.

## What Anticonsumerist Protesters Are Doing Wrong

A clear understanding of civil society and informal behavioral norms can help identify not only points of maximum leverage for changing society, but also tactics that are bound to fail. Ever since the green movement of the 1970s, the traditional strategy for trying to change consumer behavior has been through verbal preaching and admonishment. Humans love to talk, especially when we are telling other people what to do. So, we have for decades given one another vague encouragements to respect Mother Nature, consume less, recycle more, buy green, think globally, act locally, be less selfish and greedy, and live simply. In some cases, these tactics have worked surprisingly well, by creating new social norms and expectations. The preaching signals to everyone that there is a new status game in town, and that conspicuously green behavior is the best new way to display one's conscientiousness and agreeableness. In other cases, such preaching proves futile, because the sinners who most need saving (multinational corporations, military-industrial complexes) don't have personality traits, don't care about signaling them, and don't get any benefits from playing the new green game.

For example, anticonsumerism protesters often target large corporations and international trade organizations. They try to use the usual social-hominid tactics of informal social sanctioning—preaching, public humiliation, ostracism, name-calling, and throwing rotten fruit. But the objects of their wrath are faceless institutions that have no conscience or responsiveness to such sanctions, or institutional leaders and functionaries whose real social lives have no overlap with those of the protesters, and thus who are immune to suffering any real fitness costs from the protesters' disaffection. The Nestlé and WTO leaders can leave their besieged workplaces in strong, fast cars, drive to their anonymized exurban mansions, and enjoy the evening with their empathic spouses, adoring children, deferential dinner guests, and single-malt whiskies. The protesters are not their neighbors, friends, kin, colleagues, or potential lovers, so their disapproval means nothing. They are the out-group, and informal social sanctions only work within one's in-group.

The protesters would do better to aim their sticks and carrots at social in-group members who care what they think—and to recognize that their social in-group is much wider than they might realize. For typical college-student protesters, these in-groups include their likeminded, same-aged, protester-subculture friends, of course. But they also include anyone who has overlapping fitness interests by virtue of genetic relatedness, social attachment, economic codependency, spatial proximity, or repeated interaction. That is, their in-groups include all their parents, stepparents, siblings, and relatives; their housemates and neighbors; their workmates, bosses, and customers; their schoolmates and professors; their online game-playing companions, chatroom pals, and e-mail correspondents.

As adults most of us have a social network of around 150 people whom we know well enough that, if we met them in an airport, we would be happy to chat with them over drinks. In many domains, we feel comfortable praising or punishing these in-group acquaintances for good or bad behavior. We would commend them for altruism toward family, friends, children, or animals. We would condemn them—with a wince, a scowl, a gentle remonstrance, a pointed question, or an abrupt exit—if they revealed acts of cruelty or infidelity. We might even do so for ideological sins—for derogating minorities, enjoying pornography, or cursing within earshot of nuns. Yet in most developed nations, there is a strange and strong taboo against condemning in-group members for acts of conspicuous consumption. If our airport drinking buddies reveal that they have bought a new Lexus or Stanford law degree, we feel obligated to praise their success, status, and taste. If they see an ad for some new cell phone of grotesquely conspicuous precision on the airport's propaganda screens (usually tuned to CNN, in the United States), and if they comment that they covet the product, we feel reflexively inclined to assent. I wish instead that we had the guts to say something like:

> Yeah, I wanted that phone once, too. But then I thought, I already have a pretty good phone, so why do I crave this thing? It's just going to cost hours of frustration to set up, and make me stare at little electronic screens even more than I already do, so I have even fewer face-to-face conversations like the one we are having now. I think we unconsciously want these things because we want to show people that we have some attractive personal traits—things

like intelligence and conscientiousness—that lead to success as a worker and taste as a consumer. But, you know, I think these products don't even work that well to show off these traits. For instance, I can already tell that you have these traits just from talking with you for a few minutes. You make interesting, funny comments about meaningful things, so I know you're intelligent. You got through security an hour before your flight, so I know you're conscientious. Your virtues speak for themselves. We don't need to wrap all those costly goods and services around ourselves to get respect. What do you think?

Such mini-sermons might sometimes fail by seeming too direct, offensive, intimate, or weird. But they might often succeed in sparking some new thoughts and feelings, if articulated in a spirit of "Let's think through this consumerism problem together, as joint victims of bad habits" rather than "I'm a virtuous know-it-all anticonsumer, and you're a shallow, craven materialist." Especially in settings as alienating as airports, where personal identity feels paper-thin, and product branding feels thick and hot as lava, a few genuine words of personal contact and consumerist skepticism from an acquaintance can seem momentously vivid. These words might resonate in the listener's memory for weeks to come, and resound every time he sees an ad or steps into a mall; they might even be rearticulated when he meets an acquaintance of his own in some future airport bar. (Nothing ever changes for the better without someone's seeming overoptimistic about other people's thoughtfulness. . . .)

In fact, these moments of one-on-one consciousness raising, compiled across individuals, in-groups, and history, are probably the main routes by which all social change occurs. They are how civil rights, women's rights, gay rights, and animal rights got discussed and accepted. To the extent that public protests helped at all, they may have simply provided the news-feed fodder to provoke private discussions among family and friends. They were occasions for airing thoughts about topics that were previously off the radar. Once people's tacit assumptions and behavioral habits are held up to the arc lamp of thoughtful discussion, they tend to burn out like stalled film stock: flicker, scorch, bubble, whoosh. The German social philosopher Jürgen Habermas made this point already when he wrote about human emancipation through "communicative rationality" in an "ideal speech situation" within civil society—but the point bears repeating, for those who haven't curled up by the fire lately with his 1981 masterpiece *Theorie des kommunikativen Handelns: Handlungsrationalität und gesellschaftliche Rationalisierung*.

Face-to-face discussions of consumerism can often go more smoothly than confrontations about topics like racism, sexism, or homophobia. This is because when you point out that consumerism is a really inefficient way to advertise personal traits, you can praise someone's traits and tickle their vanity even as you're cluster bombing the central ideology around which they've organized their education, career, leisure, identity, status seeking, and mating strategy. As well-trained consumer narcissists, we are such insecure, praise-starved flattery sluts that a little social validation goes a long way. A friend or lover can imply that we have wasted our lives chasing consumerist dreamworlds and status mirages, as long as he or she reassures us that we still appear intelligent, attractive, and virtuous. (Don't forget to mention that, or people will cry.)

Another, more subtle way of opening such conversations is through mentioning movies that address consumerism. Most people love to talk about movies, and do so in chummy, open-minded, leisure-chat mode. (By contrast, when discussing books, magazine articles, or TV documentaries, people tend to revert to college-seminar debate mode, and become more intellectually prickly and ideologically defensive.) One can say, "You know, last night I was watching *Fight Club* on DVD again—have you ever seen it?—and I was thinking about some of its themes. . . ." Or, one could mention *American Beauty*, or *The Matrix*, or any of the other movies listed in "Further Reading and Viewing." These films have a few key features: almost every cultured person has seen at least one of them; they evoke many themes beyond consumerism, so don't elicit an instant defensive reaction, the way that *An Inconvenient Truth* or *The Corporation* would; and they offer intriguing alternative ways to display one's mental, moral, and physical traits. By starting a chat about a highly rated mainstream Hollywood film, one can slip painlessly past an acquaintance's political defenses to question consumerist assumptions and habits.

## Multiculturalism Versus Local Social Norms

There is a major legal problem with creating and enforcing new social norms in developed nations, and the problem concerns housing law. Humans are still embodied beings who interact mostly with other humans who live nearby. The social norms and trait-display tactics most favored by the local community heavily influence our behavior. However, through antidiscrimination laws regarding property rental and ownership, many countries unwittingly prohibit the development and diversification of cohesive local norms. For example, the U.S. Department of Housing and Urban Development prohibits "housing discrimination based on your race, color, national origin, religion, sex, family status, or disability." The laws were passed with the best of intentions, but they have toxic side effects on the ability of voluntarily organized communities to create the physical, social, and moral environments that their members want.

There is increasing evidence that communities with a chaotic diversity of social norms do not function very well. Some of this evidence comes from studies of ethnically diverse communities. I mention this evidence not because I think ethnic diversity is bad, but because it is one of the only proxies for social-norm diversity that has been studied so far.

For example, the political scientist Robert Putnam has found that American communities with higher levels of ethnic diversity tend to have lower levels of "social capital"—trust, altruism, cohesion, and sense of community. He and his colleagues analyzed data from thirty thousand people across forty-one U.S. communities, and found that people who live in communities with higher ethnic diversity (meaning, in the United States, more equal mixtures of black, Hispanic, white, and Asian citizens) tend to have lower:

- trust across ethnic groups
- trust within their own ethnic group
- community solidarity and cohesion
- community cooperation
- sense of political empowerment
- confidence in local government and leaders
- voter registration rates
- charity and volunteering
- investment in common goods
- interest in maintaining community facilities
- rates of carpooling
- numbers of friends
- perceived quality of life
- general happiness

These effects remained substantial even after controlling for each individual's age, sex, education, ethnicity, income, and language, and for each community's poverty rate, income inequality, crime rate, population density, mobility, and average education. Putnam did not set out to look for these effects; a great advocate of both social capital and diversity, he seems to have been appalled at these results, and published them only reluctantly. Many other researchers have reported similar findings.

I suspect that these corrosive effects of "ethnic diversity" on social capital are not really an effect of ethnicity per se, but of each ethnicity's having different social norms—different dialects, values, political attitudes, religions, social assumptions, and systems of etiquette. As Robert Kurzban and his collaborators have shown, ethnicity fades into the background when people feel motivated to cooperate with one another for the common good, based on shared interests and norms.

Communities without a coherent set of social norms just don't feel much like communities at all, so people withdraw from community life into their own families and houses.

Sadly, it has become almost impossible now for like-minded people to arrange to live together in a small community with cohesive social norms. Real norms can be sustained effectively only by selecting who moves in, by praising or punishing those who uphold or violate norms as residents, and by expelling those who repeatedly violate the norms. These are the requirements to sustain the type of cooperation called network reciprocity, in which cooperators form local "network clusters" (communities) in which they help one another. Current laws in most developed countries make network reciprocity almost impossible. Black Muslim property developers cannot set up gated communities that exclude white oppressors. Lesbians who were traumatized by childhood sexual abuse or rape cannot set up male-free zones. Pentecostals cannot exclude Satanists and Wiccans from their neighborhoods. Medical-marijuana users with cancer or glaucoma cannot set up cannabis-friendly zones. Polyamorous swingers cannot exclude monogamous puritans, or vice versa.

So, while modern multicultural communities may be very free at the level of individual lifestyle choice, they are very unfree at the level of allowing people to create and sustain distinctive local community norms and values. This is actually a bad thing, liberal ideologies notwithstanding. It means that the only way to have any influence over who your neighbors are, and how they behave, is to rent or buy at a particular price point, to achieve economic stratification. Antidiscrimination laws apply, de facto, to everything except income, with the result that we have low-income ghettos, working-class tract houses, professional exurbs: a form of assortative living by income, which correlates only moderately with intelligence and conscientiousness.

Moreover, when economic stratification is the only basis for choosing where to live, wealth becomes reified as the central form of status in every community—the lowest common denominator of human virtue, the only trait-display game in town. Since you end up living next to people who might well respect wildly different intellectual, political, social, and moral values, the only way to compete for status is through conspicuous consumption. Grow a greener lawn, buy a bigger car, add a media room. If a Pentecostal lives next to a polyamorist, the only way they can compete with each other is at the default economic level of wealth display. But if all the Pentecostals lived together, they could establish new social norms that renounce such wealth displays, and compete for status through Bible-quoting, speaking in tongues, and spreading the gospel. And if all the polyamorists lived together, they could compete for status through good conversation, great sex, minimal jealousy, maximal affection, and emotional authenticity. In both cases, their local social norms could rein in runaway consumption, and shift their time and energy to other activities that are more congruent with their most fundamental values.

This idea—the freedom to live near folks with shared values—may sound radical to members of the educated Euro-American elite, who tend to take multiculturalism, diversity, and tolerance for granted as good things. But it would sound perfectly sensible to almost any of our ancestors from any well-functioning culture in any epoch of history. It's called choosing your tribe: you have to be able to control who enters your community, and under what conditions they will be exiled. The efficiency and cohesiveness of local social life demands protection against outside threats and internal selfishness. Minimally, this requires that everyone local shares rules of etiquette for avoiding conflict, a common spoken language for resolving conflict, norms governing social, sexual, parental, kin, and economic-exchange relationships, and norms for coordinating group action, especially in emergencies. Strangely, many "communities" in developed nations lack these basic prerequisites for living together. These communities function like computers that have hardware (a physical location and infrastructure) and an operating system (a government, an economic system, and a set of metanorms concerning tolerance and diversity), but no software applications (no specific social norms governing trait display and status seeking in any domains other than wealth).

To a limited degree, people with shared values and lifestyles can sometimes coordinate their movements into particular locations. American gay men often move to San Francisco or New York. Mormons often congregate in Utah. But they are always mixed up with others hostile to their values; they must rub elbows with homophobes or atheists, and they cannot do anything about it. Under some special circumstances, people can create co-living communities with a limited set of shared rules that constrain runaway consumerism: college fraternities and sororities, communes, cooperative housing, condominium governing by internal rules and managerial boards, gated communities with restrictive covenants. However, the antidiscrimination laws still apply—these co-living systems still cannot legally select or expel members on the basis of sexual orientation or religion, which doesn't help gay men or Mormons create their own communities, and it still leaves wealth display as the default basis for social status.

So, governments should give people the freedom to create local housing communities with the power to sustain their own social norms, as long as a few basic human rights are respected. Adults must be free to move away from a community they don't like. The punishment for violating social norms must not exceed temporary ostracism or permanent exile. As John Stuart Mill argued, children must not be subject to abuse that is permanently physically or mentally disabling (such as, arguably, circumcision, clitoridectomy, religious indoctrination, or anorexia-inducing ballet lessons). Clearly, it is hard to draw the line between normal acculturation and disabling child abuse, but that has always been true, and I can't offer a panacea. Civilization progresses in part through people arguing about these issues and reaching the most enlightened, provisional, pragmatic consensus that they can achieve within their culture. At any rate, the government still has a crucial role to play in protecting the oppressed or vulnerable from the tyranny of the majority, even within the most radical of the local communities. However, if the local majority cannot impose some distinctive social norms on our forms of trait signaling, conspicuous consumption will remain the only game in town.

## Going Virtual

Apart from new ways of acquiring and displaying real physical goods, human trait display is being revolutionized by three new forms of electronic communication: mobile phones (2 billion active users globally as of mid-2008), social-networking sites such as MySpace and Facebook (each with 120 million users), and massively multiplayer online games (MMOGs) such as World of Warcraft (10 million users) and Second Life (2 million users). These are not just new forms of communication and entertainment. Viewed more broadly as part of the "Web 2.0," they are breaking down the geographical and legal barriers that have traditionally constrained people's abilities to form like-minded communities. They are allowing new virtual communities to arise with their own social norms, signaling systems, and preferred modes of trait display. They make it possible to live in a social world of one's own choosing, without regard to one's physical location.

Parents lament the time their teenagers spend with such technologies. They seem like meaningless, self-indulgent distractions from the proper role of juveniles under consumerist capitalism: (1) studying counterintuitive sciences and irrelevant humanities to display intelligence and conscientiousness, (2) working in part-time minimum-wage jobs to learn humility and even more conscientiousness, (3) participating in extracurricular activities that will look good on college and job applications, and (4) spending money on status goods, fads, and dates.

However, young people have always shown an uncanny knack for allocating their time and energy to emerging new modes of trait display that bring them the highest social and sexual payoffs. *Guanxi* is where they find it. Maybe they understand something about mobile phones, social networking, and MMOGs that older adults just don't get. Consider the historical context: every time civilizations develop new social technologies for trait display, the older generation always scoffs at the younger generation for wasting its time on the new technologies and neglecting the development of last-generation skills. The upper-class boys of ancient Greece were sometimes distracted from their proper slave-driving, olive-growing roles by that indulgent new cognitive technology for showing off their intelligence: philosophical debate at Plato's Academy. Novel reading by young Victorian women was considered a frivolous distraction from hymn singing and husband catching. For hundreds of years, higher education was a self-indulgent form of conspicuous leisure for the aristocracy and landed gentry, until aspiring bourgeois parents began to appreciate its value as an assortative mating market and an intelligence indicator for their off-spring. Beatniks talking avidly of existentialism and New Wave film in the 1960s cafés of New York and Paris were viewed as neglecting their duties to international socialism by their Old Left parents. Male hippies, seeking social status and mating opportunities through their conspicuous knowledge of Grateful Dead lyrics and Afghan hashish strains, were castigated for failing to display their fitness in the traditional macho ways: drinking, date-raping, and killing foreigners.

To young people today, mobile phones, social networking, and MMOGs are awesomely efficient ways to short-circuit consumerist conventions of trait display. Instead of spending years studying to get an educational credential, to get a high-paying job, to buy premium products, to display one's intelligence and personality traits to potential mates and friends, the kids are just displaying their traits directly through the new communication technologies. Why try to display your verbal abilities by getting a Yale degree in postmodern literary theory, when you can write your own blog? Why show off your aesthetic taste by making money to buy second-rate Impressionist paintings when you can design your own MySpace site, with your own graphics, photos, drawings, and music? Why become a pediatric doctor to show off your agreeableness, when you can just be consistently kind in your text messages? Why adopt costly religious conventions to prove your likely sexual fidelity in marriage, when you can just keep your GPS-enabled mobile phone switched on so your spouse can always call you and check your location? In every case, the new communication technology renders obsolete most traditional aspirations, values, skills, and status criteria—that is, most traditional modes of trait display.

This is confusing to the older generation, because they can never quite see how the new trait-display tactics will actually result in friends, mates, and babies. Partly this is because young people are very resourceful at inventing new dialects to hide how they communicate and interact with friends, and at arranging secret sexual liaisons. Partly it is because every generation of parents underestimates its children's capacity to find a stable mateship and economic niche as they mature in their twenties and thirties. But mainly, it is because every generation forgets how obscure and indirect its own social and sexual relationships appeared to its own parents. Nonliterate parents were perplexed that their children no longer courted in person, but wrote letters—how could mere letters lead to real relationships and real children? Early twentieth-century parents were perplexed that their kids no longer wrote letters, but talked on the phone. How could phone chatter produce grandchildren? Parents always fear that new technically mediated modes of courtship will lead children to forget how to discriminate good mates from bad mates, and that new technically mediated modes of friend making and status seeking will lead children to forget how to find a viable economic niche to feed their own kids. Yet history shows that every new generation of children has succeeded in doing both, despite the endless revolutions in technology and economic roles: hunting, gathering, herding, farming, factory work, corporate careers, credentialist professions, the electronic global economy. This track record of extreme human adaptability in socializing, mating, and parenting suggests that the coming generations will do just fine, whatever their modes of trait display.

As usual, an evolutionary perspective makes our current social concerns look smaller, more transient, and more solvable. It lets us see more clearly what is constant in human nature (the main traits that vary across people, the drive to show off these traits to others, and the drive to assess their traits), and what is variable across time and culture (verbal labels for the main traits, particular modes of trait display and trait assessment, particular forms of status and economic interchange). It reminds us that what we call "reality" today is already

90 percent social convention—our heads are already stuck most of the way up our own culture, most of the time. An evolutionary perspective gives us confidence that each new generation will find its own ways to turn new technologies into new trait-display modes and economic opportunities. It makes us aware that something else will soon replace the current system of consumerist capitalism and its key features: credentialism, workaholism, conspicuous consumption, single-family housing, fragmented kin and social networks, weak social norms, narrowly economic definitions of social progress and national status, and indirect democracy distorted by corporate interests and media conglomerates. These seemingly natural features of contemporary society will seem as alien to our great-grandchildren as mammoth hunting, field plowing, and typewriting now seem to us.

## The Grand Social Quasi Experiment

Science depends on comparative experiments to understand causation. If scientists can't assign different entities randomly to different conditions, and then measure how those conditions affect the behavior of the entities, we can't infer what really causes the behavior. Psychologists can only claim that factor $x$ "causes" some change in human behavior if we can randomly assign some people in an experimental condition with factor $x$, and some people to a condition without $x$, and see what happens. If condition $x$ reliably leads to a change in average behavior across a large enough sample of subjects, we can claim that $x$ causes the change. By contrast, the social sciences have always been crippled by the fact that they can't assign groups of people randomly to different cultures with different social norms and institutions, so they can't really infer what causes what. Even the Campbell Collaboration, a great international consortium of social scientists trying to develop the new field of evidence-based social policy, faces a real challenge in making causal claims, and advocates more randomized trials of social policy interventions in crime, education, and social welfare.

We can overcome this problem to some degree by allowing people to form a great diversity of like-minded communities, each of which has cohesive social norms. One huge benefit of legalizing such diversity is that we might learn what really makes communities work. We can compare different criteria of success across different communities and see what succeeds and what fails. The criteria can be as diverse as we like: measured levels of happiness, peacefulness, social connectedness, trait-display efficiency, cultural vibrancy, technical progress, economic well-being, environmental sustainability, whatever. The greater the number of different communities, the more diverse their norms, and the better the data about their outcomes, the faster we'll learn what kinds of social norms and institutions allow people and communities to flourish.

Of course, we still can't randomly assign people to communities; that would be unethical. But by maximizing the freedom that individuals and families have to choose their own communities, we can minimize a lot of the usual "confounds" that have made it difficult to interpret cross-cultural differences. For example, in traditional societies, a local population's culture (including social norms and institutions) is almost always confounded with the population's genetic composition, demographic structure, level of economic development, and ecological context. If one population is thriving and another is failing, it is virtually impossible to tell whether the outcome is due to differences in culture, genetics, wealth, or environment. By maximizing social and geographical mobility, we can better see the signal of culture through the noise of those confounds. So, we wouldn't have a true experimental design with true random assignment, but we'd have what behavioral scientists call a quasi-experimental design—not nearly as good, but much better than what we have available now.

To infer causality in quasi experiments, it helps to measure as many possible confounds as possible, so we can statistically control for their effects. To assess how well a community is functioning with a particular set of social norms, it would be crucial to measure the distribution of the Central Six traits among its inhabitants. Almost any community can succeed, regardless of its social norms, if it only admits highly intelligent, conscientious, agreeable, stable people. Almost any community will fail if it excludes all such people. To make fair comparisons across communities, we need to know what kind of people they contain, so we can tease out the effects of the social norms themselves.

The result would be a Grand Social Quasi Experiment. Society itself would become our laboratory, and we would learn much more effectively what makes communities succeed or fail. That knowledge would spread, and in the long run, communities would imitate the social norms, institutions, designs, and aspects of culture that work elsewhere. Successful cultural features would spread, mutate, recombine, and evolve at the social level. Failures would die out. This would lead to an especially fast and effective form of "cultural group selection," as the anthropologists Robert Boyd and Peter Richerson call it. One nice thing about cultural group selection is that better cultures can spread from population to population without the people themselves colonizing or killing one another. Another nice thing about cultural group selection is that we do not have to pretend that we already know what kinds of social norms, institutions, and designs work. We don't have to do social engineering that is centralized, compulsory, or arrogant, like the grand experiments of twentieth-century totalitarianism. We can be humble, let people experiment with their social arrangements, observe the results, and embrace what works. In other words, we can recognize that cultural evolution, like biological evolution, is much smarter than we are.

# 17

## Legalizing Freedom

CHANGING INFORMAL SOCIAL NORMS is usually the most effective way to change human behavior. This can often be accomplished without any change in government policies, programs, taxes, or laws. But sometimes, existing features of formal government impose perverse incentives that limit people's freedom to change old social norms—especially the norms that govern social and sexual signaling. This chapter considers some ways that citizens could encourage their governments to change the nature of consumerism—not by outlawing conspicuous consumption, but by allowing its many alternatives to flourish more easily. That is, government policies don't need to actively discourage conspicuous consumption. They just need to stop unwittingly (or intentionally) promoting conspicuous consumption by constraining human choices and relationships.

One problem with most current governments is that they prioritize economic growth (as mismeasured by GDP per capita) over citizens' happiness, quality of life, efficiency of trait display, and breadth and depth of social networks. The latter outcomes are not actually any harder to measure than GDP per capita. For example, the UN Human Development Index (HDI) measures overall quality of life fairly well by taking into account life expectancy, literacy, and educational attainment; this index puts Iceland, Norway, Australia, and Canada at the top, and the Democratic Republic of the Congo at the bottom. Even the World Bank now encourages countries to measure the value of "ecosystem services," "natural capital," "social capital," education, and knowledge. Yet even the most comprehensive measures of life quality still tend to overlook some key variables, such as average number of hours worked per year (eighteen hundred for the United States and

Japan versus fourteen hundred for Germany, France, Norway, and the Netherlands).

However, these nonmonetary outcomes just don't produce tax revenue to support government bureaucracies and politicians' own vainglorious trait displays, such as hospitals named after ministers, and aircraft carriers pimped out as victory-speech platforms. So, Bhutan remains the only nation on earth to prioritize explicitly Gross National Happiness over Gross Domestic Product. Since governments have incentives to maximize tax revenue, they also have incentives to bias the kinds of trait displays that their citizens use toward those that require paid employment and consumer spending. So, if you want to understand how political power biases the trait-display systems of our society, you have to track the money on which politics depends: taxation to support the government, and campaign finance to support the individual politicians. These are the key points of democratic leverage where citizens can enact policy changes that free everybody to display their traits in more diverse and more accurate ways.

## From Income Taxes to Consumption Taxes

You may think that tax policy sounds like the most boring topic in the world. That is precisely what most governments, corporations, and special interests would like you to think, because tax policy is where much of society and the economy gets shaped. It is also where well-informed citizens can achieve socioeconomic revolutions with astonishing speed and effectiveness—but only if they realize how much power they might wield in this domain. If citizens don't understand taxes, they don't understand how, when, and where their government expropriates money, time, and freedom from their lives. They also don't understand how most governments bias consumption over savings, and bias some forms of consumption over other forms, thereby distorting the trait-display systems that people might otherwise favor.

One strange aspect of taxation is the pervasive mismatch between what governments do and what experts recommend. In most developed countries, including the United States and the United Kingdom, governments rely mostly on income taxes. However, most people who have thought hard about tax policy—ranging from conservative economists to eco-activists—favor consumption taxes. The sales tax levied by most U.S. states is a simple example of a consumption tax: consumers pay an extra 4–8 percent on top of most retail purchases, which goes to the state government. The value-added tax (VAT) favored in Europe is a slightly more complex consumption tax that is levied not just on retail purchases by individual consumers, but on all transactions throughout the supply chain, from mining and manufacturing through distribution and retail.

Virtually all economists agree that consumption taxes, compared with income taxes, encourage less consuming but more earning, saving, investing, and charitable giving. By making consumption relatively more expensive, consumption taxes make all these other activities relatively less expensive, so people tend to shift their money from consumption to the other activities. Thus, consumption taxes tend to reduce conspicuous consumption and promote longer-term retirement security, family wealth, social welfare, technical progress, and economic growth. In essence, income taxes penalize people for what they contribute to society (labor and capital), whereas consumption taxes penalize people for what they take out of society (new retail purchases). So, to tax experts, it is no surprise that U.S. and U.K. citizens spend too much and don't save enough, relative to what would be optimal for society and even for themselves. The problem is not just that consumers lack moral self-restraint at the individual level, but that citizens face perverse economic incentives at the aggregate level.

Most Eastern European countries (such as Ukraine, Latvia, Serbia, and Romania) already switched to consumption taxes in the 1990s, with largely positive results. Most Western European countries, plus Japan, already levy some sort of VAT that functions as a consumption tax. A U.S. consumption tax has been advocated by the economists Alan Greenspan and Robert Frank, and by the FairTax organization.

FairTax suggests a simple, visible, federal retail sales tax (of about

23 percent) on all retail purchases of new goods or services by individual consumers at the point and time of sale. Its proposal would abolish all federal personal and corporate income taxes, and all gift, estate, capital gains, alternative minimum, Social Security, Medicare, and self-employment taxes. It would be a mildly progressive tax, giving a "prebate" to ensure that no one below the poverty level pays the tax. It would also raise revenue more fairly from illegal immigrants and tourists (who do not file tax returns but do buy things), and underground-economy workers (such as burglars, prostitutes, and babysitters, all of whom typically underreport their income but who cannot hide their retail purchases).

More conservative analysts tend to favor a flat tax—a fixed percent of the yearly difference between one's income and one's consumer spending (excluding all savings, investments, and charitable donations). Either the FairTax or the flat tax would be simple to implement and easy to understand, and would save hundreds of billions of dollars in tax-compliance costs.

The economist Robert Frank has suggested a more steeply progressive consumption tax. This would be administered like the flat tax, but the percent paid would increase with each individual's net consumption, minus a moderate deduction for basic living costs (around $30,000 for a family of four). The consumption tax would start at a low rate such as 10 percent, but would gradually climb as a family's net consumption increased, to a top marginal rate of about 100 percent for families who spend more than $10 million per year. Robert Frank's reasoning, like mine, is that many purchases function as positional goods that display one's wealth, status, or personality traits rather than yielding true happiness benefits or fitness payoffs to the purchaser. ("True happiness" here has the uncontentious meaning of "subjective well-being" or "overall life satisfaction," as it can be measured reliably and validly with many different questionnaires.) For big spenders seeking high status through consumption, even a very high consumption tax rate would not actually decrease their happiness or their fitness. It would just redirect some of their money from zero-sum forms of conspicuous waste, precision, and reputation into positive-sum forms of government spending. All my arguments are highly supportive of Robert Frank's proposal for a progressive consumption tax.

## Different Consumption Tax Rates for Different Products?

For consumption taxes collected at the place and time of retail sale, it would be easy to charge different tax rates on different kinds of goods and services. It might be fair and reasonable to impose a higher consumption tax rate on products that impose higher "negative externalities" (costly side effects) on society and the environment, so government programs can offset those side effects. For example, if a $1 pack of cigarettes imposes about $6 of additional costs on a country's nationalized health care system, to cover the expenses of caring for the increased number of lung cancer and emphysema cases, then the consumption tax on cigarettes should be 600 percent. (Currently, U.S. states impose a consumption tax to cover the higher health-care costs for smokers, but this tax ranges from only $.18 per twenty-cigarette pack in Mississippi to $2.58 in New Jersey.). Such a tax could reduce the long-term health-care costs of smoking, especially in large poor countries, such as China (where 65 percent of men—420 million—smoke) and India (where 45 percent of men—250 million—smoke). That's 1.3 billion lungs regularly exposed to carcinogens.

As another example, if a $1 pineapple is transported two thousand miles by truck on federal highways, and imposes $2 of wear and tear on the nation's transport infrastructure, its consumption tax rate should be 200 percent—much higher than for a locally grown apple or avocado. Conversely, socially and ecologically innocuous products (such as bicycles, universities, and iTunes downloads) might pay much lower consumption tax rates.

Before barcodes, laser scanners, and computerized inventory systems, it would have been impractical for retailers to charge product-specific consumption tax rates. Now each product type could be assigned to a certain consumption tax rate by federal government economists, on the basis of real data concerning externalities, and this tax rate would automatically be imposed as cashiers scan the item or charge for the service.

Any products that yield "positive externalities" (whose acquisition and use actually saves resources for government, society, and the environment) might be subject to a negative consumption tax rate (effectively, a government subsidy). These positive-externality products might include home insulation (to minimize global warming), airbags retrofitted to older cars (to minimize costly injuries), and vocational training (to minimize unemployment).

At the other extreme, the consumption tax rate should be very, very high for any products that impose massive negative externalities. Consider handgun ammunition. Currently, one can buy five hundred rounds of 9 mm ammunition for about $110 from online U.S. retailers—about twenty-two cents each. But each round of ammunition has a slight chance of falling into the wrong hands and killing someone. How slight? About 10 billion rounds are sold per year in the United States. There are about thirty thousand gun-related deaths in the United States per year (including suicides, homicides, and accidents). Assuming the typical gun death involves one round of ammo, the chance that any given round will end up killing someone is about thirty thousand divided by 10 billion, or three per million. Now, a person's life is generally reckoned to be worth about $3 million, according to the usual cost-benefit-risk analyses by highway engineers, airlines, and hospitals. If each bullet has a three per million chance of negating a $3 million life, then that bullet imposes an expected average cost on society of $9. That's about forty times its conventional retail cost of $0.22, so, by my reasoning, it should be subject to a consumption tax rate of 4,000 percent. This is obviously a rough calculation; it ignores the injury costs of nonlethal shootings (which would increase the tax) and the crime-deterrence effects, if any, of citizens having ammo (which would decrease the tax). In any case, the five-hundred-round box of ammo should cost about $4,500, not $110.

Such a high tax rate would in no way undermine the sacred Second

Amendment, which the National Rifle Association holds so dear. People could still have the right to keep and bear arms, but would simply have to pay society the true expected costs that their dangerous hobby imposes on others. They could pay $4,500—more than one month's median U.S. household income—for a five-hundred-round box of ammo for "recreational shooting." If they're unwilling to pay that much, they should offer a principled argument for why the rest of us should subsidize the aggregate social, economic, medical, and funeral costs of their avocation.

Similar arguments might apply to motorcycles and pit bulls (both of which maim and kill their owners at ghastly rates). The consumption tax rate should also be high for service industries that impose costly side effects on society in the form of obesity (fast-food restaurants), noise pollution (concrete demolition), or false hope that delays medical or psychiatric treatment (purveyors of Christian Science, Scientology, homeopathy, or vibrating phytoplankton). Yes, people should be free to choose the level of physical risk they wish to impose on themselves, but they should pay steeply for the likely social costs if the risks are very high.

By this reasoning, a product-specific consumption tax would allow governments to legalize a much wider range of goods and services. Most products that are illegal somewhere (drugs, prostitution, transfatty acids) are outlawed because regulators assume that they impose unacceptable externalities on society (addiction, AIDS, obesity). If evidence-based consumption taxes negated the true costs of these externalities, the main economic and moral reason for outlawing such products would evaporate. The world's most influential regulators (who are now in Brussels rather than Washington) could focus on minimizing harm from the truly dangerous products and business practices.

## Drawing the True Cost Map

It would present a great challenge to determine the appropriate consumption tax rate for each product category, and to do so based on legitimate empirical research rather than ideology. For many product categories, governments and environmental activists might try to maximize the applicable consumption tax rate, while marketers would try to minimize it. At first, only the most obvious social costs—health care, transport, pollution—could be estimated in dollar-per-product terms. Quantifying the average long-term health-care costs to the user of smoking a pack of cigarettes would be much easier than quantifying the costs of the secondhand smoke. Quantifying the highway-repair costs of transporting each pineapple two thousand miles would be much easier than quantifying the alleged patriarchy-reinforcing, rape-promoting effects of each pornographic magazine.

A vast amount of new scientific research would need to be done on product-specific externalities, and that research would have to be very carefully funded, analyzed, and interpreted. There would be ubiquitous confounds and complexities—maybe fatter people tend to smoke more, and the fat, rather than the smoke, is contributing to lung cancer. Maybe low-agreeableness males tend to buy porn more often, and also commit more rapes, but perhaps the porn doesn't cause the rapes. Ideally, the externality measurements would use the same sort of randomized trials now used to determine the safety and efficacy of clinical drugs. This would require a radical change in the government's mind-set—from arrogantly assuming that we know how policies will influence behavior to humbly realizing that we need to run social experiments to determine their effects.

For instance, if we really want to know whether soda consumption increases obesity and diabetes rates, we could run a study using the consumption tax itself. The soda consumption tax rate could be set at a wide range of different random levels (so that a can of soda could cost anywhere from $.25 to $50) in several thousand different towns and cities. Given sufficient policing to minimize the resulting black market in untaxed soda, higher tax rates would reduce soda sales, so would lead to different soda consumption rates in different areas. By recording those consumption rates and obesity and diabetes rates in each area for several years, we could see exactly how much each can of soda increases the average obesity and diabetes rates, and resulting health care costs. The soda consumption tax rate could then be set accordingly. It may sound coldhearted to run such studies, which in effect cast soda drinkers as guinea pigs. However, this is exactly the data needed to settle the debate that has been raging since the 1970s between anti-soda activists and soda manufacturers. Both groups, insofar as they have any intellectual integrity, should want better causal data on the issue, and the only way to obtain that data is through this sort of randomized study. Once a few such studies across different countries are published, the issue would be resolved.

Skeptics might object that to set appropriate product-specific consumption tax rates would require a vast new government bureaucracy. We would need thousands of economists, statisticians, actuaries, and psychologists to measure all the externalities, risks, and costs of every product class. That is true, but that is precisely what we need: good solid data about the true social and environmental costs of the goods and services we buy. If we don't collect and analyze such data, all arguments about the social and environmental effects of different policies are just blather. They can't be evidence-based if there is no evidence.

As a government project, a massive scientific initiative to gather evidence about consumption externalities doesn't sound as exciting as an Apollo program to send a man to the moon. However, it might give us a much greater long-term payoff: an accurate map of how our consumer decisions affect our society and environment. We could call it something catchy, like the True Cost Map, rather than something bureaucratic, like the Evidence-Based International Reference Matrix of Product-Specific Externalities (EBIRMOPSE for short?). Developing the True Cost Map would have many other benefits, such as giving meaningful work to thousands of underemployed social scientists who would otherwise run around teaching the Wrong Radical Model to gullible undergraduates.

Garrett Hardin's "tragedy of the commons" arises whenever people don't have to pay for the true externalities that their consumption choices impose on others. However, when people are responsible for such payments, their consumption choices become aligned almost perfectly with the interests of society at large. If fish consumers had to pay the true full price that their fish eating imposes on endangered stocks, overfishing would be automatically reduced. Ideally, the consumption tax on each class of products should be used to offset the specific externalities imposed by that class of product. In reducing the tragedy of the commons effect, it doesn't really matter whether the consumption tax dollars are collected and burned, or collected and spent on useful social and environmental programs. The latter seems more sensible though.

To conservatives and libertarians, such a consumption-tax system may sound like a nightmarish new way for the government to interfere with the free market. In practice, it would severely restrict the government's power to meddle. Suppose product-specific tax rates were required to be based on rigorous empirical evidence concerning externalities. Then governments could only impose higher-than-minimum rates on products that demonstrably impose costs on other people and on their environment. Even the most ardent libertarians, such as Robert Nozick, have recognized the need for a "night watchman state" that protects people's lives and property from other people's bad behavior. Laws against murder and robbery are, from an economist's viewpoint, simply ways of deterring people from imposing negative externalities such as death or property loss on others. Product-specific consumption tax rates would simply deter consumers from imposing other sorts of negative externalities (such as pollution, highway wear and tear, and risks of getting shot) on others.

All negative externalities are, by definition, encroachments on other people's lives and property. So, even hard-core libertarians who believe that governments should do nothing more than protect people from such encroachments should be willing to accept a consumption tax specifically designed to counteract such encroachments. From this viewpoint, the consumption tax is not paternalistic meddling. Rather, it is a classical "Pigovian tax" designed to correct the negative externalities of market activity. Democratic governments are, among other things, ways for people to manage the externalities of human economic behavior in free markets. From that perspective, it seems reasonable that governments should impose consumption taxes designed to neutralize each product type's externalities. In other words, we should be free to choose what we buy and how we live, as long as we pay the fair price for every harm we do to others in the process.

## Promoting Product Longevity

Consumption taxes could help solve another problem with modern consumerism. At the moment, corporations can often maximize long-run sales through planned obsolescence and shoddy build quality. If you're a mop maker, you get a fourfold-higher revenue stream from selling plastic-handled mops that cost $8 and break within two years (yielding $4 per year in sales per customer), rather than steel-handled mops that cost $12 and last twelve years (yielding $1 per year). The environmental costs of making each mop might be similar—perhaps each imposes $4 of externalities. But the plastic-handled mops need replacing six times as often, so over a mop user's lifetime, the use of plastic-handled mops will impose those $4 externalities six times as often. The consumption tax rate imposed on a product category could be scaled up for products of lower build quality that depreciate faster, break faster, and need replacement more often. For example, steel-handled mops might pay only one-sixth the consumption tax rate of plastic-handled mops—perhaps a rate of 12 percent rather than 72 percent. Thus, the steel-handled mop's total price would be $12 plus 12 percent, or $13.44, whereas the plastic-handled mop's total price would be $8 plus 60 percent, or $13.76. The sensible consumer would then buy the steel-handled mop because it's cheaper, even if he wasn't aware of the different build qualities, average breakage rates, and expected replacement rates.

These depreciation-sensitive consumption tax rates could apply to much costlier goods that impose much heavier externalities, and the benefits to society and the environment would be even greater.

For instance, consider two midsize SUVs that cost about $29,000: the Toyota Highlander and the Ford Explorer. Suppose for the sake of argument that the Highlander is designed to run for an average of 240,000 miles before it wears out, whereas the Explorer is designed to run for an average of only 120,000 miles. It might make sense to charge twice the consumption tax rate on the Explorer (maybe 40 percent instead of 20 percent), because habitual Explorer buyers will have to buy new vehicles twice as often as habitual Highlander buyers. Consumers would then face a choice between a $34,800 Highlander ($29,000 plus 20 percent) and a $40,600 Explorer ($29,000 plus 40 percent). This would make reliability differences more salient to consumers, and would shift buying patterns from more-disposable products to longer-lasting ones.

Depreciation-sensitive rates would be even more important and beneficial in the housing market. In 1997 we bought a house in Redhill, England, that was built in 1898. It had solid brick walls, thick roof tiles, and sturdy woodwork, and was apparently designed to last about two hundred years. In 2001 we moved to Albuquerque and shopped for houses. The 1950s first-generation suburban houses were fairly well built (cinder-block walls, oak floors, coved plaster ceilings), so we bought one of those. However, the new houses all seemed to be built from chopsticks and cardboard: walls of half-inch gypsum board nailed onto widely-spaced two-by-fours, floors of nylon carpet above oriented-strand board, flat roofs of tar paper and grit, thinly sprayed stucco siding, hollow-core doors, and plastic bathtubs. They are apparently designed to last about ten years, and everybody knows it, which is why they are sold as "used" houses after five years of all-too-visible depreciation. If consumption tax rates on new houses were much lower for better-built houses with greater expected longevity, such rickety abominations would never be built. Developers might not use solid brick walls as they did in 1898 England, but they might invent something even longer lasting. We could take our inspiration from the Clock of the Long Now project, which aims to build a mechanical clock on a Nevada mountaintop that will keep time for ten thousand years. We want developers to build houses that our grandchildren could inherit, so they need not spend their lives paying a mortgage on yet another generation of misconceived shacks. We might even consider arranging economic incentives so we can enjoy built environments that age gracefully through hundreds of years, like Umbrian villas or Oxford rectories. It's the least we can do for future generations.

Good design can minimize depreciation even for more transient goods, such as clothes. The mainstream fashion industry intentionally produces clothes with extreme and quirky designs and colors so they can be rendered aesthetically obsolete more easily the following year, when the new design features are decided at Paris Fashion Week, and the new colors are "forecast" by the Inter-Society Color Council and other trendsetters. By contrast, the new trend for "eco-fashion" by designers such as Rebecca Earley ("Green is the new black") aims to create attractive clothes that are both physically and aesthetically durable. They are produced from sustainable, nonpolluting sources (organic cotton, rayon, or hemp rather than conventionally grown cotton, which requires high pesticide use), and are designed to resist and hide stains from spills, dirt, and sweat, and to require lower-energy washing, faster drying, and no ironing. Eco-fashion uses more classically appealing designs and colors that cannot be rendered obsolete as easily, and so arguably should be subject to a lower consumption tax than extreme high fashion of the sort that dominates *Vogue* and *In Style*.

It would be a bit tricky to implement a consumption tax that was both differential (based on product-specific externalities and depreciation rates) and progressive (based on the purchaser's annual total consumption). The differential consumption tax (with different rates for different products) is collected most easily at the point and time of sale by the retailer. By contrast, the progressive consumption tax is collected most easily if people file an annual tax return that reports the difference between their total income and their total savings and charitable giving. Some ingenuity would be required to develop a practical, secure method of collecting a differential, progressive consumption tax at the point of retail sale. However, if we can invent international computerized systems for managing cell-phone calls, airline reservations, debit charges, and Amazon.com shipments, it seems likely that we could organize such a consumption tax system. The technical challenges would be solvable, and the social and environmental benefits would be dramatic.

## What the Consumption Tax Might Accomplish

Suppose we followed the experts' advice and derived all government revenue from a consumption tax averaging about 25 percent on all new retail purchases. How would this shift our consumer behavior and our social systems of trait display? I can see five major benefits.

First, people would reduce, reuse, and recycle more, because only new retail purchases would be taxed. There would be no consumption tax on used items sold through the secondary market, such as used cars, secondhand books, thrift-store clothing, antique furniture, or pre-owned houses. If you wanted a new-built house in a developer's new exurban gated community, you would have to pay the consumption tax on the full price of the house; whereas if you wanted to renovate an older house in an established city or town, you would pay the consumption tax only on the building materials and labor costs. If you wanted the newest model of car, you would pay the consumption tax on the car's full price, but if you were content with a used car that just needed some new tires, you would pay tax only on the tires. This would create incentives to make the best use of what previous generations have already built and produced. It would hugely expand the secondary market, and the scale and efficiency with which used goods get traded and used. Estate sales, yard sales, eBay, classified ads, and flea markets would flourish. The matter and energy that is invested in the production of physical objects would stay in circulation longer, and their ecological costs would be amortized over a much longer period.

Second, the consumption tax would also create incentives for people to buy longer-lasting goods that have a higher resale value in the secondary market. They would be motivated to maintain and repair those goods more conscientiously, and to replace them less often. If the old washing machine can be repaired with a new $20 part (plus $5 in tax) rather than replaced with a new washing machine for $500 (plus $125 in tax), people might opt for the repair more often, and see more clearly the absurdity of premature replacement. We already saw how depreciation-sensitive consumption tax rates would increase manufacturers' incentives to produce longer-lasting products. These other effects—consumers thinking ahead about product resale values and repairability—would allow retailers to charge more for products that are easier to maintain, repair, upgrade, and redecorate, and so would further encourage the production of such items.

Third, the consumption tax would encourage people to buy products that consume less energy and matter to operate. As gas, oil, charcoal, electricity, batteries, water, printer ink cartridges, and coffee filters would all be subject to the full consumption tax, consumers might prefer cars that get better mileage and printer cartridges that use ink more efficiently. The promotion of green buildings would be especially important. About 45 percent of the world's energy budget is used to heat and cool buildings (far more than to power all vehicles combined), but new materials and design methods can easily reduce building energy needs by about 80–90 percent. Currently, such energy efficiency is rarely maximized in new construction, because it costs about 15 percent more than energy-inefficient methods. However, as consumers realize how much they could save in highly taxed running costs by spending a tiny bit more up front, the profitability and popularity of energy-efficient building should grow. Or, energy-efficient buildings could simply be subject to a lower consumption tax rate.

Fourth, the consumption tax would promote social capital and neighborly camaraderie. It would apply only to formal retail purchases, so it would push people to barter, borrow, and make more things themselves. Citizens would develop and use less-formal systems of trade, reciprocity, reputation, and trust. Friends would have higher incentives to lend one another useful items, and to make one another more practical gifts. Neighbors would have higher incentives to trade mutually useful services—babysitting, gardening, tutoring, carpooling, housepainting, barn raising—to avoid the consumption tax premium for formally hired employees. To economists who believe that an atomistic division of labor in a formal job market necessarily yields the highest economic efficiency, this sort of informal service sector might sound horribly misconceived. But to most humans—members of a species evolved to flourish in small-scale societies built on mutual recognition, respect, and trust—it might sound rather pleasant and fulfilling.

Finally, the consumption tax would increase savings, investment, and charity. In effect, each dollar that an individual allocated to investment or charity would buy 25 percent more than each dollar allocated to consumption. Investment and charity could be made more salient in everyday life as signals of one's personal traits, through various new electronic databases, displays, and technologies. Then conspicuous consumption could easily fall from favor as the standard trait-display system. More matter, energy, time, and skill would be invested in the long-term infrastructure of civilization, and less in burning through short-term displays of conspicuous waste, precision, and reputation.

## The Will to Change

Civilizations change most dramatically when their status-signaling systems change. Marx overlooked an important truth: the means of display, not just the means of production, are crucial factors in economic and social revolutions. Signaling systems show strong lock-in effects: once a signaling convention such as runaway consumerism gets established, it can be very hard for a population to shift to another set of conventions. The signaling conventions start to look like an inevitable outcome of cosmic evolution, rather than a system of historically defined cultural norms. Conspicuous consumerism is neither natural nor inevitable, but just one possible mode of human trait display.

Also, because conspicuous consumerism requires an alienating suppression of many natural social instincts, it is especially poor at arming us with good indicators of some key personality traits such as agreeableness and conscientiousness. To display these traits, consumers are slowly shifting from traditional conspicuous consumption to conspicuously ethical consumption. Highly agreeable and conscientious consumers now try to buy recycled computer-printer paper at a 40 percent premium above virgin paper, or purchase shade-grown Fair Trade coffee at $12 a pound rather than Wal-Mart generic coffee at $3 a pound. Conspicuously ethical consumption is certainly one potent way to improve the world, but there are even more revolutionary possibilities. If we could find more efficient ways for people to display their Central Six traits—especially agreeableness and conscientiousness—the social and environmental results could be even more positive.

The last few chapters have outlined several ideas for shifting our trait-display system away from conspicuous consumption and toward other signaling games that are more natural, humane, efficient, accurate, environmentally responsible, and fulfilling. The ideas may seem radical, especially to Americans, who are unaccustomed to really significant changes in socioeconomic policy. Other countries better appreciate the potential speed and depth of change. In the past twenty years Europe has been transformed from a patchwork of ragged empire remnants into the world's largest, richest, best-integrated economy. China, India, and Russia have abandoned socialist planning in favor of free markets and free trade. The number of countries with multiparty democracy has increased from less than fifty to more than eighty. There are now 6.7 billion people on earth. Half of them are under thirty—they had not yet reached puberty when Bill Clinton was elected United States president, when Tony Blair was elected U.K. prime minister, or when Kurt Cobain killed himself.

In the global long run, it doesn't much matter how the United States or U.K. change their consumption patterns, because their populations and economies are such a small and shrinking proportion of the entire world's. The two countries that will matter most are the ones with vast populations and fast-growing economies, namely China (1.3 billion people, 10 percent economic growth per year) and India (1.1 billion people, 8 percent growth per year). Those two countries still have a good chance to shift their trait-display systems before conspicuous consumption becomes locked in as the cultural norm.

Dramatic policy changes are much easier for China to accomplish, because its stronger central government is less constrained by the competing special-interest groups (ethnicities, religions, castes, languages) that characterize the feverish Indian democracy. For example, from 1995 to 2007, China built about thirty thousand miles of expressways—about the same in a twelve-year period as the United States had built in the forty years after the Federal-Aid Highway Act was enacted in 1956. The current "fourth-generation" Chinese leaders are mostly bright, practical-minded engineers, already keenly aware of the environmental and social costs of conspicuous consumption. Most of my key ideas—social norms favoring status through informal assessment of people's personal traits and moral virtues, better objective psychometric testing for education and employment, a progressive, product-specific consumption tax—would not undermine their political power or China's social stability. Indeed, many of them dovetail nicely with traditional Confucian ideals (respect for merit and virtue within individuals, families, and states) and practices (such as the imperial civil service exams), and with communist principles (at least, at the abstract level of promoting the positive externalities and minimizing the negative externalities of social and economic activity). They also fit well with China's strong new interest in building eco-sustainable New Urbanist cities, such as Tangye New Town in Jinan Province. If China adopts such farsighted policies, and thrives, it will serve as an exciting new role model for other developing and developed nations.

Despite India's more chaotic political system, it already shows thoughtful ambivalence about the idea of economic growth for its own sake, and an increasing determination not to mimic the worst aspects of American-style conspicuous consumption. The older generation of

Indian intellectuals still clings to the Nehru-era socialist ideals that
imposed the lamentable "Hindu rate of economic growth" and delayed
India's climb out of mass poverty from the 1940s through the early
1990s. However, those socialist ideals also undercut the reputability of conspicuous consumption—as do the older traditions of Hindu
asceticism, Buddhist nonattachment, and Muslim brotherhood. Many
of the younger generation of Indian entrepreneurs obtained degrees
and work experience in the United States and Europe, and are fluent
in English, the international language of both business and science.
These younger entrepreneurs have seen both the social and cultural
costs of runaway consumerism, and the ways that some countries have
minimized those costs. Many of them seem to be setting a good example by leapfrogging over conspicuous consumption straight to conspicuously ethical investment and charity, in just a single generation. It is
crucial for India's younger generation to avoid viewing runaway consumerism as the only "modern" alternative to the stultifying traditions
of religion, caste, patriarchy, socialism, and identity politics.

## Why the Sky Won't Fall If We Gently Shift
## Our Signaling Systems

The changes suggested in the last few chapters might sound alarming to
some. If people suddenly adopted product-specific consumption taxes,
self-policing communities of like-minded citizens, and these other
innovations, wouldn't some very bad things happen? If people bought
fewer goods and services, then consumer demand would shrink, sales
volumes and prices would drop, unsold inventories would increase,
supply chains would clog, companies would go bankrupt, factories and
malls would shut, and unemployment would skyrocket. Then currencies, equities, bonds, and real estate values would go into free fall, and
investment bankers would fling themselves from tall buildings in New
York, London, and Hong Kong. Tax revenues—whether from income,
consumption, or investments—would evaporate. First recession, then
depression, then economic collapse. The next steps are predictable:

governments fall, police vanish, riots flare, anarchy spreads, services collapse, utilities fail, people starve, babies cry, gangs rape, plagues spread, warlords rise, armies clash, nukes explode. The misery would be long enough to cause hysteria, and wide enough to cause great pain. Without conspicuous consumption to keep everyone busy, wouldn't we all end up living in an eco-communo-primitivist dystopia whether we wanted to or not? Wouldn't we go straight from *American Beauty* through *American Psycho* to *Mad Max*?

Such fears would be reasonable if the changes happened too quickly for markets, companies, and governments to adapt. Moon-size meteors are bad for biodiversity, because they kill too many species too quickly for evolutionary adaptation to compensate. (Remember the Permian-Triassic extinction 251 million years ago that destroyed 96 percent of all sea species and 70 percent of all land vertebrates? Oh, maybe not.) Likewise, short sharp shocks to the social norms governing our trait-display systems can be bad for human societies—consider the 1789 French Revolution (outcome: Napoléon Bonaparte), 1917 Russian Revolution (outcome: Joseph Stalin), and 1994 Republican Revolution (outcome: Newt Gingrich).

But over the longer term, economies are astonishingly resilient. Joseph Schumpeter observed that economies actually thrive on the "creative destruction" sparked by radical changes in technology and society. Other Austrian and Chicago School economists have emphasized that free markets are the most ingenious, resourceful, and adaptive systems ever invented by humans. Free markets embody the aggregate intelligence of every buyer, seller, and innovator, and, like life itself, they always find a way forward. This isn't wishful thinking, it's historical fact. When hired shepherds were replaced by border collies, they become factory workers. Buggy-whip makers morph into bumper sticker designers. Medieval knights are rendered obsolete by the crossbow, but their male descendants may become Formula 1 drivers, navy pilots, or Segway enthusiasts. Milkmaids are replaced by vacuum pulsation milking machines, but their female descendants may become software project managers or radio astronomers. The

Civil War ends cotton-plantation slavery, but the South rises again by inventing Nascar and the Atlanta hip-hop scene. In each case, our ancestors could hardly have imagined the jobs people hold now, or the products they buy, or the ways they seek status.

Whole sectors of the economy rise and fall, but the market lives on. Our ancestors survived vast economic disruptions without going extinct, adapting from hunter-gatherers to farmers and herders in the Neolithic Revolution, from peasants to factory workers in the Industrial Revolution, and from thing makers to status purveyors and experience providers in the marketing revolution. Given historical precedent, the same creative destruction is likely to yield viable business, investment, and employment opportunities even if conspicuous consumption is replaced almost entirely by ethical consumption and ethical investment. We just have to preserve the institutional and cultural prerequisites for free markets to work: peace, the rule of law, property rights, stable currency, efficient regulation, honest government, and social norms of truth, trust, fairness, and honor. I've heard that these prerequisites already exist in Hong Kong, Silicon Valley, and Switzerland, so they're clearly not unattainable.

## Conclusion: Self-Gilding Genes

Consumerist capitalism is largely an exercise in gilding the lily. We take wondrously adaptive capacities for human self-display—language, intelligence, kindness, creativity, and beauty—and then forget how to use them in making friends, attracting mates, and gaining prestige. Instead we rely on goods and services acquired through education, work, and consumption to advertise our personal traits to others. These costly signals are mostly redundant or misleading, so others usually ignore them. They prefer to judge us through natural face-to-face interaction. We think our gilding dazzles them, though we ignore their own gilding when choosing our own friends and mates.

This is an absurd way to live, but it's never too late to come away from it. We can find better ways to combine the best features of prehistoric human life and modern life. Eco-communo-primitivism alone offers little more than squalor, ignorance, and boredom. Runaway consumerism alone offers little more than narcissism, exhaustion, and alienation. We need the freedom to explore different ways of displaying our traits to the people we care about. We need to legalize more diverse forms of trait display. We need to switch from an income tax that promotes short-term runaway consumption to a consumption tax that promotes longer-term ethical investment, charity, social capital, and neighborly warmth.

Above all, we need the freedom to live assortatively with like-minded people, and to establish a much wider variety of local communities with their own values and norms, sustained by their own forms of praise and punishment. Some communities may retain the focus on conspicuously amoral consumption. Some may shift to conspicuously ethical consumption or conspicuously ethical investment. I hope that eventually, most will invent systems of human trait display that we can't even imagine now. When people are free to move where they want, live as they want, and allocate their time, energy, and money to the trait displays they want, they will discover some fabulous new ways to live and display. Humans may never give up their drives for status, respect, prestige, sexual attractiveness, and social popularity, but these drives can be channeled to yield a much higher quality of life than runaway consumerism offers. We can flaunt our fitness with more individuality, ingenuity, and enlightenment.

# Exercises for the Reader

## The Natural-Living Test

This quantifies how closely your life matches that of our happier ancestors. Write down honestly how many times in the past month you have had each of the experiences below:

- Rocked a newborn baby to sleep
- Made up a story and told it to a child
- Felt the sunrise warm your face
- Satisfied a genuine hunger by eating ripe fruit
- Satisfied a genuine thirst by drinking cool water
- Shown courage in protecting a child from danger
- Shown leadership and resourcefulness in an emergency
- Shared a meal with parents, siblings, or other close relatives
- Gossiped with an old friend
- Made a new friend
- Made something beautiful and gave it to someone
- Repaired something that was broken
- Improved a skill through diligent practice
- Learned something new about a plant or animal that lives near you
- Changed your mind about something important on the basis of new evidence
- Followed good advice from someone older
- Taught a useful skill, charming art, or interesting fact to someone younger
- Petted a furry animal such as a dog, cat, or monkey
- Worked with earth, clay, stone, wood, or fiber

- Comforted someone dying
- Walked over a hill and across a stream
- Identified a bird by its song
- Played a significant role in a local ritual, festival, drama, or party
- Played a team sport
- Made a physical effort to achieve a collective goal with others
- Sustained silent eye contact with someone to show affection
- Shamed someone who was behaving badly, for the greater good
- Resolved a serious argument using humor, emotional self-control, and social empathy
- Sang, danced, or played instruments with a group of friends
- Made friends laugh out loud
- Reached a world-melting mutual orgasm with a sexual partner
- Experienced sublime beauty that made your hair stand on end
- Experienced an oceanic sense of oneness with the cosmos that made you think, *This is how church should feel*
- Applied the Golden Rule by helping someone in need
- Warmed yourself by an open fire under stars

Now, add up all the numbers that you wrote for each item above. If your total score is lower than 100 and you do not feel as happy as you would like, write a five-hundred-word essay explaining why you expect your life to be happy or meaningful if you are not doing anything meaningful for others or feeding your brain any of the natural experiences that it evolved to value and to find meaningful.

## *The Mall Visit*

First, clear your mind by spending two hours in a flotation tank, sensory-deprivation chamber, or meditation ritual. Then go to a large local shopping mall, leaving all your cash, checks, debit cards, and credit cars at home. Spend two hours at the mall as a primatologist would, observing and taking notes on shoppers as hypersocial, semi-monogamous, status-seeking primates.

## The Possessions Exercise

List the ten most expensive things (products, services, or experiences) that you have ever paid for (including houses, cars, university degrees, marriage ceremonies, divorce settlements, and taxes). Then, list the ten items that you have ever bought that gave you the most happiness. Count how many items appear on both lists.

## The Home-Archaeologist Exercise

Walk around your home for an hour. Note ten of your physical possessions that could last long enough, in some form, to be discovered in five thousand years by a future archaeologist. For each product, list the practical and signaling functions that the archaeologist might attribute to the possession. Compare those inferred functions with your conscious reasons for buying the possession.

## The Sims Exercise

Play The Sims 2 for a couple of weeks, and consider whether your life as a consumer has any more meaning than that of your Sims.

## The Estate-Sale Exercise

On a weekend morning, go to three consecutive estate sales (sales of dead people's possessions from their now-unoccupied houses). Watch the buyers swarm over the flotsam looking for bargains. Examine the items and contemplate how mute the once-prestigious goods have become, now that their owners can no longer use, display, or talk about them.

## The Sex-Differences Exercise

If you are a woman, buy a *Maxim* men's magazine. If you are a man, buy a *Cosmopolitan* magazine. Read the entire thing—every word—both editorial content and advertising, cover to cover, in the course of a week. List ten ways in which the mind of the opposite sex is similar to your mind, and ten ways in which it is a freakish alien planet beyond all comprehension.

### The Central-Six Exercise

As honestly as possible, estimate your own percentile values on the Central Six traits (intelligence, openness, conscientiousness, agreeableness, stability, extraversion). A percentile of 10 percent means only 10 percent of people would score lower on that trait than you; a percentile of 90 percent means than 90 percent of people score lower than you. Then estimate these trait percentiles for each of these people: your three most recent sexual partners, three closest friends, three closest co-workers, and three nearest neighbors. For each person and each trait, calculate the percentile difference between you and them. See if you feel closest to those who are most similar to you on the Central Six.

### The Car-Personality Correction

Write down the make, model, year, and features of whatever vehicle you own and drive most often. Analyze its brand personality by estimating the percentile scores on the Central Six traits that it seems to be conveying through its design and advertising. Compare your vehicle's personality profile to yours, and note any discrepancies. Add two bumper stickers that might correct those discrepancies by signaling your personality traits more accurately.

### The Holiday Challenge

Next holiday, make all the gifts yourself that you give to friends and family. Compare their gratitude to previous years in which you bought all the gifts.

## The Social-Norm Exercise

List five forms of conspicuous consumption that you most detest, and that you would most like to see outlawed. Then cross out each one that people would not pursue if doing so made them subject to public humiliation, satire, and ridicule, even if they remained legal. If any items are left uncrossed-out on the list, write to your elected representative about them. If all are crossed out, consider how to incorporate more humiliation, satire, and ridicule into your daily life.

## The Externalities Exercise

List the three most recent things you bought that cost more than $50. For each, try to estimate what the product-specific consumption tax should have been on the item, based on the total negative externalities that the item imposes on other people and the environment. Give those estimated amounts to your favorite charity.

## The Neighborhood-Selection Exercise

Imagine that you are a young, single, childless adult just graduating from college. Imagine that your country already contained a million or so different neighborhoods that select their residents to fit almost any conceivable combination of individual traits, based on age, sexual orientation, education, occupation, ethnicity, religion, politics, hobbies, interests, physical appearance, Central Six traits, and preferences with regard to lifestyle, activities, social norms, sexual norms, local amenities, and architectural and landscape aesthetics. Write down the ten criteria you would use to select a neighborhood into which you would want to move, to start your working life. Then write down how much more rent you would be willing to pay per month to live in such a community, compared with the place you actually lived as a young adult.

# Further Reading and Viewing

## Non-fiction

Topics below are listed roughly in their order of appearance in the book.

References are in standard American Psychological Association format, except that I've also included each first author's first name. For works with more than two authors, only the first author's name is given, followed by 'et al.' Abbreviations: "J." = "Journal", "NY" = New York", "U." = "University."

For readers who want more detail and citations for particular facts, findings, and ideas, this book's website (provisionally www.geoffrey-miller.com) includes a complete set of over 1,200 endnotes and over 1,600 references. Whenever facts, books, people, products, or organizations are not specifically noted, details can be easily accessed from the Web by searching the relevant name through Google.com or Wikipedia.org. Except where otherwise noted in those endnotes, all quotations in the book are from Robert Fitzhenry (1993) *The Harper Book of Quotations*.

### The Nature of Capitalism and Consumerism

Bowles, Samuel, et al. (2005). *Understanding Capitalism*. NY: Oxford U. Press.

De Soto, Hernando (2003). *The Mystery of Capital*. NY: Basic Books.

Galbraith, John K. (1952). *American Capitalism*. NY: Houghton Mifflin.

Hart, Stuart L. (2007). *Capitalism at the Crossroads*. Upper Saddle River, NJ: Wharton School Publishing.

Lindblom, Charles E. (2002). *The Market System*. New Haven, CT: Yale U. Press.

Muller, Jerry Z. (2003). *The Mind and the Market*. NY: Anchor.

Reich, Robert B. (2007). *Supercapitalism*.NY: Knopf.

## The Psychology of Happiness Versus Runaway Consumerism

Bruni, Luigino, & Porta, P. L. (Ed.). (2007). *Handbook on the Economics of Happiness.*Northampton, MA: Edward Elgar.

Easterbrook, Gregg (2004). *The Progress Paradox*. NY: Random House.

Frank, Robert H. (2000). *Luxury Fever*. Princeton, NJ: Princeton U. Press.

Frey, Bruno S. (2008). *Happiness*. Cambridge, MA: MIT Press.

Gilbert, Daniel T. (2006). *Stumbling on Happiness*. NY: Knopf.

Kasser, Tim (2002). *The High Price of Materialism*. Cambridge, MA: MIT Press.

Kasser, Tim, & Kanner, A. D. (2004). *Psychology and Consumer Culture*. Washington, DC: American Psychological Association.

Lane, Robert E. (2000). *The Loss of Happiness in Market Democracies*. New Haven, CT: Yale U. Press.

Nussbaum, Martha C., & Sen, A. (Eds.). (1993). *The Quality of Life*. NY: Oxford U. Press.

Scitovsky, Tibor (1992). *The Joyless Economy*. Oxford, UK: Oxford University Press.

Weiner, Eric (2008). *The Geography of Bliss*. NY: Twelve.

## Consumerism and Capitalism in Historical Context

Acemoglu, Daron, & Robinson, J. A. (2006). *Economic Origins of Dictatorship and Democracy*. NY: Cambridge U. Press.

Bernstein, William J. (2008). *A Splendid Exchange*. NY: Atlantic Monthly Press.

Carrier, James G. (2006). *A Handbook of Economic Anthropology*. Northampton, MA: Edward Elgar.

Cohen, Lizabeth (2003). *A Consumer's Republic*. NY: Alfred A. Knopf.

Clark, Gregory (2007). *A Farewell to Alms*. Princeton, NJ: Princeton U. Press.

Collins, Robert M. (2002). *More*. NY: Oxford U. Press.

Cross, Gary (2000). *An All-consuming Century*. NY: Columbia U. Press.

Davidson, James (1997). *Courtesans and Fishcakes*. NY: St. Martins Press.

Earle, Timothy (2002). *Bronze Age Economics*. Boulder, CO: Westview Press.

Fogel, Robert W. (2004). *The Escape from Hunger and Premature Death, 1700–2100*. NY: Cambridge U. Press.

Frank, Thomas (1997). *The Conquest of Cool*. Chicago, IL: U. Chicago Press.

Galbraith, John K. (1958). *The Affluent Society*. Boston: Houghton Mifflin.

Jardine, Lisa (1998). *Worldly Goods*. NY: W. W. Norton.

Johnson, Allen, & Earle, T. (2000). *The Evolution of Human Societies*. Stanford, CA: Stanford U. Press.

Landes, David S. (1999). *The Wealth and Poverty of Nations*. NY: W. W. Norton.

Leach, William (1993). *Land of Desire*. NY: Pantheon Books.

Sale, Kirkpatrick (2006). *After Eden*. Durham, NC: Duke University Press.

Schama, Simon (1997). *The Embarrassment of Riches*. NY: Vintage.

Schor, Juliet B., & Holt, D. (Eds.). (2000). *The Consumer Society Reader*. NY: New Press.

Wilk, Richard R., & Cliggett, L. (2004). *Economies and Cultures*. Boulder, CO: Westview Press.

Twitchell, James B. (1999). *Lead us Into Temptation*. NY: Columbia U. Press.

### Globalization and Consumerism Across Cultures

Bhagwati, Jagdish (2007). *In Defense of Globalization*. NY: Oxford U. Press.

Friedman, Thomas L. (2000). *The Lexus and the Olive Tree*. NY: Anchor.

Friedman, Thomas L. (2005). *The World is Flat*. NY: Farrar, Straus, and Giroux.

Levinson, Marc (2006). *The Box*. Princeton, NJ: Princeton U. Press.

McMillan, John (2003). *Reinventing the Bazaar*. NY: W. W. Norton.

Sachs, Jeffrey (2006). *The End of Poverty*. NY: Penguin.

Sassen, Saskia (2006). *Territory, Authority, Rights*. Princeton, NJ: Princeton U. Press.

Sen, Amartya (2000). *Development as Freedom*. NY: Knopf.

Shiller, Robert J. (2003). *The New Financial Order*. Princeton, NJ: Princeton U. Press.

Singer, Peter (2004). *One World*. New Haven, CT: Yale U. Press.

Stiglitz, Joseph E. (2002). *Globalization and its Discontents*. London: Penguin Books.

### Evolutionary Biology

Carroll, Sean B. (2006). *The Making of the Fittest*. NY: W. W. Norton.

Dawkins, Richard (2006). *The Selfish Gene*. Oxford, UK: Oxford U. Press.

Nowak, Martin A. (2006). *Evolutionary Dynamics*. Cambridge, MA: Belknap Press.

Ridley, Mark (2003). *Evolution*. London: Blackwell.

Wilson, David S. (2007). *Evolution for Everyone*. NY: Delacorte Press.

### Animal Behavior (Especially Primates and Parasites)

Alcock, John (2005). *Animal Behavior*. Sunderland, MA: Sinauer.

Combes, Claude (2005). *The Art of Being a Parasite*. Chicago: U. Chicago Press.

Maestripieri, Dario (Ed.). (2005). *Primate Psychology*. Cambridge, MA: Harvard U. Press.

Nelson, Randy J. (2005). *An Introduction to Behavioral Endocrinology*. Sunderland, MA: Sinauer.

Poulin, Robert (2006). *Evolutionary Ecology of Parasites*. Princeton, NJ: Princeton U. Press.

Strier, Karen B. (2005). *Primate Behavioral Ecology*. NY: Allyn & Bacon.


### Human Evolution

Boyd, Robert, & Silk, J. (2005). *How Humans Evolved*. NY: Norton.

Dunbar, Robin I. M. (2005). *The Human Story*. London: Faber & Faber.

Jobling, Mark A., et al. (2003). *Human Evolutionary Genetics*. NY: Garland Science.

Sawyer, G. J., et al. (2007). *The Last Human*. New Haven, CT: Yale U. Press.


### Evolutionary Psychology

Buss, David M. (Ed.). (2005). *The Handbook of Evolutionary Psychology*. NY: Wiley.

Buss, David M. (2008). *Evolutionary Psychology*. NY: Allyn & Bacon.

Cartwright, John (2008). *Evolution and Human Behavior*. Cambridge, MA: MIT Press.

Crawford, Charles, & Krebs, D. (Eds.). (2008). *Foundations of Evolutionary Psychology*. NY: Taylor & Francis.

Dunbar, Robin I. M., & Barrett, L. (Eds.). (2007). *The Oxford Handbook of Evolutionary Psychology*. NY: Oxford U. Press.

Gangestad, Steven W., & Simpson, J. (Eds.). (2007). *The Evolution of Mind*. NY: Guildford Press.

Kenrick, Douglas T., & Luce, C. L. (Eds.). (2004). *The Functional Mind*. NY: Pearson.

Pinker, Steven (2002). *The Blank Slate*. NY: Viking.

Platek, Steven M., et al. (Eds.). (2006). *Evolutionary Cognitive Neuroscience*. Cambridge, MA: MIT Press.
Segerstråle, Ullica C. O. (2001). *Defenders of the Truth*. NY: Oxford U. Press.


## Consumerism in Evolutionary Context

Conniff, Richard (2002). *The Natural History of the Rich*. NY: W. W. Norton.
Frank, Robert H. (2007). *The Economic Naturalist*. NY: Basic Books.
Gandolfi, Arthur E., et al. (2002). *Economics as an Evolutionary Science*. Piscataway, NJ: Transaction.
Neiva, Eduardo (2007). *Communication Games*. NY: Mouton de Gruyter.
Saad, Gad (2007). *The Evolutionary Bases of Consumption*. Mahwah, NJ: Lawrence Erlbaum.
Seabright, Paul (2005). *The Company of Strangers*. Princeton, NJ: Princeton U. Press.
Shermer, Michael (2007). *The Mind of the Market*. NY: Times Books.


## The Evolutionary Psychology of Certain Kinds of Products

Flesch, William (2008). *Comeuppance*. Cambridge, MA: Harvard U. Press. [fiction]
Gottschall, Jonathan, & Wilson, D. S. (Eds.). (2005). *The Literary Animal*. Evanston, IL: Northwestern U. Press. [fiction]
Hersey, George L. (1999). *The Monumental Impulse*. Cambridge, MA: MIT Press. [architecture]
Pollan, Michael (2007). *The Omnivore's Dilemma*. NY: Penguin. [food]
Salmon, Catherine, & Symons, D. (2001). *Warrior Lovers*. London: Weidenfeld & Nicholson. [fiction]
Wansink, Brian (2006). *Mindless Eating*. NY: Bantam. [food]


## Sexual Selection, Sexual Evolution, and Sex Differences

Baron-Cohen, Simon (2004). *The Essential Difference*. London: Penguin.
Buss, David M. (2003). *The Evolution of Desire*. NY: Free Press.
Campbell, Anne (2002). *A Mind of her Own*. NY: Oxford U. Press.
Ellis, Lee (2008). *Sex Differences*. NY: Psychology Press.
Ellison, Peter T. (2003). *On Fertile Ground*. Cambridge, MA: Harvard U. Press.

Judson, Olivia (2002). *Dr. Tatiana's Sex Advice to all Creation*. NY: Owl Books.
Kauth, Michael E. (Ed.). (2007). *Handbook of the Evolution of Human Sexuality*. NY: Routledge.
Mealey, Linda (2000). *Sex Differences*. San Diego, CA: Academic Press.
Thornhill, Randy, & Gangestad, S. W. (2008). *The Evolutionary Biology of Human Female Sexuality*. NY: Oxford U. Press.


## Social Critiques of Runaway Consumerism and Workaholism

Bell, Daniel (1996). *The Cultural Contradictions of Capitalism*. NY: Basic Books.
Conley, Lucas (2008). *Obsessive Branding Disorder*. NY: PublicAffairs.
De Graaf, John, et al. (2005). *Affluenza*. San Francisco, CA: Berrett-Koehler.
Ehrenreich, Barbara (2001). *Nickel and Dimed*. NY: Henry Holt.
Frank, Thomas, & Weiland, M. (Eds.) (1997). *Commodify your Dissent*. NY: W. W. Norton.
Gini, Al (2000). *My Job, my Self*. NY: Routledge.
Hirsch, Fred (1995). *Social Limits to Growth*. Cambridge, MA: Harvard U. Press.
Hochschild, Arlie (2003). *The Managed Heart*. Berkeley, CA: U. California Press.
Klein, Naomi (2002). *No Logo*. NY: Picador.
Klein, Naomi (2008). *The Shock Doctrine*. NY: Picador.
Lasch, Christopher (1991). *The Culture of Narcissism*. NY: W. W. Norton.
Lasn, Kalle (2000). *Culture Jam*. NY: Harper.
Locke, John L. (1999). *The De-voicing of Society*. NY: Simon & Schuster.
Marcuse, Herbert (1956). *Eros and Civilization*. London: Routledge & Kegan Paul.
Marcuse, Herbert (1964). *One-dimensional Man*. Boston: Beacon Press.
Packard, Vance (1959). *The Status Seekers*. NY: David McKay.
Packard, Vance (1960). *The Waste Makers*. NY: Van Rees Press.
Remnick, David (Ed.). (2001). *The New Gilded Age*. NY: The Modern Library.
Schor, Juliet B. (1992). *The Overworked American*. NY: Basic Books.
Schor, Juliet B. (1998). *The Overspent American*. NY: Basic Books.
Taylor, Mark C. (2004). *Confidence Games*. Chicago, IL: U. Chicago Press.
Whybrow, Peter C. (2006). *American Mania*. NY: W. W. Norton.

## Environmental Critiques of Runaway Consumerism

Ayres, Robert U., & Ayres, L. W. (2002). *A Handbook of Industrial Ecology*. Brookfield, VT: Edward Elgar.

Borgerhoff Mulder, Monique, & Coppolillo, P. (2005). *Conservation*. Princeton, NJ: Princeton U. Press.

Brown, Lester R. (2008). *Plan B 3.0*. NY: W. W. Norton.

Diamond, Jared (2005). *Collapse*. NY: Penguin.

Hayden, Andres (2000). *Sharing the Work, Sparing the Planet*. London: Zed Books.

McDonough, William, & Braungart, M. (2002). *Cradle to Cradle*. NY: North Point Press.

McKibben, Bill (2007). *Deep Economy*. NY: Times Books.

Myers, Norman, & Kent, Jennifer (2004). *The New Consumers*. Washington, DC: Island Press.

Papanek, Victor (1995). *The Green Imperative*. NY: Thames and Hudson.

Rosenzweig, Michael (2003). *Win-win Ecology*. NY: Oxford U. Press.

Sale, Kirkpatrick (2007). *Human Scale*. Buffalo, NY: BlazeVOX Books.

Slade, Giles (2007). *Made to Break*. Cambridge, MA: Harvard U. Press.

Strasser, Susan (1999). *Waste and Want*. NY: Henry Holt & Co.

Weyler, Rex (2004). *Greenpeace*. Emmaus, PA: Rodale.


## Corporate Power, Media Influence, and Lobbyists

Bakan, Joel (2004). *The Corporation*. NY: Free Press.

Frank, Thomas (2000). *One Market Under God*. NY: Doubleday.

Heinz, John P., et al. (1997). *The Hollow Core*. Cambridge, MA: Harvard U. Press.

Korten, David C. (2001). *When Corporations Rule the World*. West Hartford, CT: Kumarian Publishers.

Nestle, Marion (2002). *Food Politics*. Berkeley, CA: U. California Press.

Palast, Greg (2004). *The Best Democracy Money Can Buy*. NY: Plume.

Postman, Neil (2005). *Amusing Ourselves to Death*.NY: Penguin.

Rushkoff, Douglas (1999). *Coercion*. NY: Riverhead Books.

Schlosser, Eric (2001). *Fast Food Nation*. NY: Harper Perennial.

Tye, Larry (1998). *The Father of Spin*. NY: Crown.

## Evolution, Development, and Childhood

Bjorklund, David F., & Pelligrini, A. D. (2002). *The Origins of Human Nature.* Washington, DC: American Psychological Association.

Bloom, Paul (2005). *Descartes' Baby.* NY: Basic Books.

Ellis, Bruce J., & Bjorklund, D. F. (Eds.). (2005). *Origins of the Social Mind.* NY: Guilford Press.

Harris, Judith R. (1998). *The Nurture Assumption.* NY: Free Press.

Hewlett, Barry S., & Lamb, M. E. (Eds.). (2005). *Hunter-gatherer Childhoods.* New Brunswick, NJ: Aldine Transaction.

Hrdy, Sarah B. (1999). *Mother Nature.* NY: Ballantine.

Salmon, Catherine, & Shackelford, T. K. (Eds.). (2008). *Family Relationships.* NY: Oxford U. Press.

Weisfeld, Glenn (1999). *Evolutionary Principles of Human Adolescence.* NY: Basic Books.

## Children as Consumers, Consumer Socialization

Linn, Susan (2004). *Consuming Kids.* NY: New Press.

Louv, Richard (2008). *Last Child in the Woods.* Chapel Hill, NC: Algonquin.

McNeal, James U. (2007). *On Becoming a Consumer.* Boston: Butterworth-Heinemann.

Schor, Juliet B. (2004). *Born to Buy.* NY: Scribner.

## Consumer Behavior, Marketing, Advertising, and Branding in General

Aaker, David A. (2007). *Strategic Market Management.* NY: Wiley.

Baker, Michael J. (2005). *The Marketing Book.* Boston: Butterworth-Heinemann.

Batey, Mark (2008). *Brand Meaning.* NY: Routledge.

Buttle, Francis (2008). *Customer Relationship Management.* Boston: Butterworth-Heinemann.

Gerzema, John, & Lear, E. (2008). *The Brand Bubble.* San Francisco, CA: Jossey-Bass.

Hawkins, Del I., Best, R. J., & Coney, K. E. (2003). *Consumer Behavior.* NY: McGraw-Hill.

Kotler, Philip, & Armstrong, G. (2007). *Principles of Marketing.* Upper Saddle River, NJ: Pearson Prentice Hall.

Levitt, Theodore (1983). *The Marketing Imagination*. NY: Macmillan.

Neumeier, Marty (2005). *The Brand Gap*. Indianapolis, IN: New Riders/Hayden.

Schmitt, Bernd, & Simonson, A. (1997). *Marketing Aesthetics*. NY: Free Press.

Twitchell, James B. (2005). *Branded Nation*. NY: Simon & Schuster.

Wells, William D., et al. (2005). *Advertising*. NY: Prentice-Hall.

Walker, Rob (2008). *Buying In*. NY: Random House.

Weitz, Barton A., & Wensley, R. (Eds.). (2002). *Handbook of Marketing*. NY: Sage.

Zaltman, Gerald, & Zaltman, L. H. (2008). *Marketing Metaphoria*. Boston, MA: Harvard Business School Press.

## The Psychology of Economics and Consumer Decision-Making

Ariely, Dan (2008). *Predictably Irrational*. NY: HarperCollins.

Becker, Gary S. (1998). *Accounting for Tastes*. Cambridge, MA: Harvard U. Press.

Camerer, Colin (2003). *Behavioral Game Theory*. Princeton, NJ: Princeton U. Press.

Frank, Robert H. (2007). *Microeconomics and Behavior*. NY: McGraw-Hill.

Frey, Bruno S., & Stutzer, A. (Eds.). (2007). *Economics & Psychology*. Cambridge, MA: MIT Press.

Gigerenzer, Gerd (2007). *Gut Feelings*. NY: Viking.

Gladwell, Malcolm (2007). *Blink*. NY: Back Bay Books.

Glimcher, Paul W., et al. (Eds.). (2008). *Neuroeconomics*. NY: Academic Press.

Harford, Tim (2008). *The Logic of Life*. NY: Random House.

Haugvedt, Curtis P. et al. (2008). *Handbook of Consumer Psychology*. NY: Routledge.

Levitt, Steven D., & Dubner, S. J. (2005). *Freakonomics*. NY: William Morrow.

Schwartz, Barry (2004). *The Paradox of Choice*. NY: HarperCollins.

Shiller, Robert J. (2001). *Irrational Exuberance*. NY: Broadway.

Underhill, Paco (1999). *Why we Buy*. NY: Simon & Schuster.

## Emotions and Motivations Relevant to Consumer Behaviour

Evans, Dylan, & Cruse, P. (2004). *Emotion, Evolution, and Rationality*. Oxford, UK: Oxford U. Press.

Gagnier, Regina (2001). *The Insatiability of Human Wants*. Chicago, IL: U. Chicago Press.

Lewis, Alan (Ed.). (2008). *The Cambridge Handbook of Psychology and Economic Behavior*. NY: Cambridge U. Press.
Lewis, Michael, et al. (Eds.). (2008). *Handbook of Emotions*. NY: Guilford Press.
Nesse, Randolph M. (Ed.). (2001). *Evolution and the Capacity for Commitment*. NY: Russell Sage.
Panksepp, Jaak (2004). *Affective Neuroscience*. NY: Oxford U. Press.

## Cultural Analyses of Consumerism, Marketing, Advertising, and Media in General

Baudrillard, Jean (1998). *The Consumer Society*. NY: Sage.
Benjamin, Walter (2002). *The Arcades Project*. Cambridge, MA: Belknap Press.
Bloom, Paul N., & Gundlach, G. T. (Eds.). (2000). *Handbook of Marketing and Society*. NY: Sage.
Bourdieu, Pierre (1984). *Distinction*. Cambridge, MA: Harvard U. Press.
Brooks, David (2000). *Bobos in Paradise*. NY: Simon & Schuster.
Chomsky, Noam (2002). *Manufacturing Consent*. NY: Pantheon.
Cowen, Tyler (1998). *In Praise of Commercial Culture*. Cambridge, MA: Harvard U. Press.
Deleuze, Gilles, & Guattari, F. (1987). *A Thousand Plateaus*. Minneapolis, MN: U. Minnesota Press.
Douglas, Mary, & Isherwood, B. (1980). *The World of Goods*. London, UK: Penguin.
Ewan, Stuart (1999). *All-consuming Images*. NY: Basic Books.
Frank, Robert L. (2007). *Richistan*. NY: Crown Books.
Graeber, David (2001). *Toward an Anthropological Theory of Value*. NY: Palgrave Macmillan.

## Cultural Analyses of Certain Kinds of Products

Agins, Teri (2000). *The End of Fashion*. NY: Harper. [clothing]
Battelle, John (2006). *The Search*. NY: Portfolio. [Google and search engines]
Belk, Russell W. (2001). *Collecting in a Consumer Society*. London: Routledge.
Bradsher, Keith (2002). *High and Mighty*. NY: Public Affairs. [SUVs]
Chaplin, Heather, & Ruby, A. (2006). *Smartbomb*. NY: Algonquin Books. [videogames]
Clark, Taylor (2007). *Starbucked*. Boston: Little, Brown, & Co. [coffee]

Denton, Sally, & Morris, R. (2001). *The Money and the Power*. NY: Knopf. [gambling]

Glasmeier, Amy K. (2000). *Manufacturing Time*. London: Guilford. [watches]

Blumenthal, Howard J., & Goodenough, O. R. (2006). *This Business of Television*. NY: Billboard Books.

Hart, Matthew (2001). *Diamond*. NY: Walker & Co.

Horst, Heather, & Miller, D. (2006). *The Cell Phone*. NY: Berg.

Illouz, Eva (1997). *Consuming the Romantic Utopia*. Berkeley, CA: U. California Press. [courtship products]

Johnson, Steven (2005). *Everything Bad is Good for You*. NY: Riverhead Books.

Kahney, Leander (2006). *The Cult of Mac*. San Francisco, CA: No Starch Press. [computers]

Kushner, David (2004). *Masters of Doom*. NY: Random House. [computer games]

Levy, Steven (2007). *The Perfect Thing*. NY: Simon & Schuster. [iPods]

Ling, Rich (2004). *The Mobile Connection*. San Francisco, CA: Morgan Kaufmann.

Manning, Robert D. (2000). *Credit Card Nation*. NY: Basic Books.

Mead, Rebecca (2007). *One Perfect Day*. NY: Penguin. [weddings]

Radosh, Daniel (2008). *Rapture Ready!* NY: Scribner. [Christian products]

Reichert, Tom, & Lambiase, J. (Eds.). (2006). *Sex in Consumer Culture*. Mahwah, NJ: Lawrence Erlbaum Associates.

Seagrave. Kerry (2002). *Vending Machines*. Jefferson, NC: McFarland.

Singh, Simon, & Ernst, E. (2008). *Trick or Treatment*. NY: W. W. Norton. [alternative medicine]

Steiner, Wendy (2001). *Venus in Exile*. NY: Free Press. [modern art]

Velthius, Olav (2005). *Talking Prices*. Princeton, NJ: Princeton U. Press. [contemporary art]

### *Costly Signaling Theory, Mutations, and Fitness Indicators*

Bradbury, Jack W., & Vehrencamp, S. L. (1998). *Principles of Animal Communication*. Sunderland, MA: Sinauer Associates.

Hauser, Marc D., & Konishi, M. (Eds.). (2004). *The Design of Animal Communication*. Cambridge, MA: MIT Press.

Maynard Smith, John, & Harper. D. (2004). *Animal Signals*. NY: Oxford U. Press.

Ridley, Mark (2001). *The Cooperative Gene*. NY: Free Press.

Searcy, William A., & Nowicki, S. (2005). *The Evolution of Animal Communication*. Princeton, NJ: Princeton U. Press.

Smith, David L. (2007). *Why we Lie*. NY: St. Martin's Press.

Zahavi, Amotz, & Zahavi, A. (1997). *The Handicap Principle*. NY: Oxford University Press.

## The Nature and Psychology of Status

De Botton, Alain (2004). *Status Anxiety*. NY: Penguin.

Frank, Robert H. (1985). *Choosing the Right Pond*. Oxford, UK: Oxford U. Press.

Lareau, Annette, & Conley, D. (Eds.). (2008). *Social Class*. NY: Russell Sage Foundation.

Seabrook, John (2001). *Nobrow*. NY: Vintage.

Silverstein, Michael J. (2006). *Treasure Hunt*. NY: Portfolio.

Silverstein, Michael J., & Fiske, N. (2003). *Trading Up*. NY: Penguin.

Thomas, Dana (2007). *Deluxe*. NY: Penguin.

## Conspicuous Consumption, Positional Goods, and Inequality

English, James F. (2005). *The Economy of Prestige*. Cambridge, MA: Harvard U. Press.

Frank, Robert H., & Cook, P. J. (1995). *The Winner-take-society*. NY: Free Press.

James, Oliver (2008). *The Selfish Capitalist*. London: Vermillion.

Rothkopf, David (2008). *Superclass*. NY: Farrar, Straus, & Giroux.

Veblen, Thorstein (1899). *The Theory of the Leisure Class*. NY: Macmillan.

## Alternatives to Consumerism: Voluntary Simplicity, Thrift

Elgin, Duane (1993). *Voluntary Simplicity*. NY: Quill.

Fox, Nichols (2002). *Against the Machine*. Washington, DC: Island Press.

Hawken, Paul (2008). *Blessed Unrest*. NY: Penguin.

HonorÈ, Carl (2005). *In Praise of Slowness*. NY: HarperOne.

Luhrs, Janet (1997). *The Simple Living Guide*. NY: Broadway.

Merkel, Jim (2003). *Radical Simplicity*. Gabriola Island, British Columbia, Canada: New Society Publishers.

Stanley, Thomas (2000). *The Millionaire Mind*. Kansas City, KS: Andrews McMeel Publishing.

## Evolution of Culture and Society

Baumeister, Roy F. (2005). *The Cultural Animal: Human Nature, Meaning, and Social Life*. NY: Oxford U. Press.

Barkow, Jerome H. (2005). *Missing the Revolution*. NY: Oxford U. Press.

Boyd, Robert, & Richerson, P. J. (2005). *The Origin and Evolution of Cultures*. Oxford, UK: Oxford U. Press.

Schaller, Mark, & Crandall, C. S. (Eds.). (2004). *The Psychological Foundations of Culture*. Mahwah, NJ: Erlbaum.

Sperber, Dan (1996). *Explaining Culture*. Oxford, UK: Blackwell.


## Psychology of Culture, Fads, Fashions, and Memes

Blackmore, Susan (1999). *The Meme Machine*. Oxford, UK: Oxford U. Press.

Heath, Chip, & Heath, D. (2007). *Made to Stick*. NY: Random House.

Shennan, Stephen (2002). *Genes, Memes, and Human History*. London: Thames & Hudson.

Surowiecki, James (2005). *The Wisdom of Crowds*. NY: Anchor.


## Person Perception and Social Cognition

Ambady, Nalini, & Skowronski, J. J. (Eds.). (2008). *First Impressions*. NY: Guilford Press.

Cacioppo, John T., et al. (Eds.). (2006). *Social Neuroscience*. Cambridge, MA: MIT Press.

Cialdini, Robert (2008). *Influence*. Boston, MA: Allyn & Bacon.

Dalrymple, Theodore (2007). *In Praise of Prejudice*. NY: Encounter Books.

Fiske, Susan T., & Taylor, S. E. (2008). *Social Cognition*. NY: McGraw-Hill.

Fletcher, Garth J. O. (2002). *The New Science of Intimate Relationships*. Oxford, UK: Blackwell.

Funder, David C. (1999). *Personality Judgment*. NY: Academic Press.

Gosling, Samuel D. (2008). *Snoop*. NY: Basic Books.

Kenrick, Douglas T., et al. (2005). *Social Psychology*. NY: Allyn & Bacon.

Schaller, Mark, et al. (Eds.). (2006). *Evolution and Social Psychology*. NY: Psychology Press.

## The Web, Social Networking, Online Games, and New Media

Boellstorff, Tom (2008). *Coming of Age in Second Life*. Princeton, NJ: Princeton U. Press.

Castronova, Edward (2005). *Synthetic Worlds*. Chicago, IL: U. Chicago Press.

Gillin, Paul (2007). *The New Influencers*. Sanger, CA: Quill Driver Books.

Kelly, Richard V. (2004). *Massively Multiplayer Online Role-playing Games*. Jefferson, NC: McFarland & Co.

Kirby, Justin, & Marsden, P. (Eds.). (2005). *Connected Marketing*. London: Butterworth-Heinemann.

Li, Charlene, & Bernoff, J. (2008). *Groundswell*. Boston, MA: Harvard Business School Press.

Mansell, Robin, et al. (Eds.). (2007). *The Oxford Handbook of Information and Communication Technologies*. NY: Oxford U. Press.

Meadows, Mark S. (2008). *I, Avatar*. Indianapolis, IN: New Riders Press.

Tapscott, Don, & Williams, A. D. (2008). *Wikinomics*. NY: Portfolio.

Weber, Larry (2007). *Marketing to the Social Web*. NY: Wiley.


## Physical Attractiveness, Beauty Culture, and Physical Fitness Indicators

Cooper, Pamela (1998). *The American Marathon*. Syracuse, NY: Syracuse U. Press.

Etcoff, Nancy (1999). *Survival of the Prettiest*. NY: Doubleday.

Farrell-Beck, Jane, & Gau, C. (2002). *Uplift*. Philadelphia, PA: U. Pennsylvania Press.

Hersey, George L. (1996). *The Evolution of Allure*. Cambridge, MA: MIT Press.

Jablonsky, Nina G. (2006). *Skin*. Berkeley, CA: U. California Press.

Kolata, Gina B. (2003). *Ultimate Fitness*. NY: Farrar, Straus, & Giroux.

Kuczynski, Alex (2006). *Beauty Junkies*. NY: Doubleday.

Morris, Desmond (1985). *Bodywatching*. NY: Crown.

Peiss, Kathy (1998). *Hope in a Jar*. NY: Henry Holt & Co.

Rhodes, Gillian, & Zebrowitz, L. A. (Eds.). (2001). *Facial Attractiveness*. Westport, CT: Ablex.

Rothman, Sheila M., & Rothman, D. J. (2003). *The Pursuit of Perfection*. NY: Pantheon.

Stearns, Stephen C., & Koella, J. C. (2008). *Evolution in Health and Disease*. NY: Oxford U. Press.

Voland, Eckart, & Grammer, K. (Eds.). (2003). *Evolutionary Aesthetics*. Berlin: Springer.

## *Mental Fitness Indicators: Language, Art, Music, Humor, Creativity*

Dutton, Denis (2008). *The Art Instinct*. London: Bloomsbury Press.
Eysenck, Hans J. (1995). *Genius*. Cambridge, UK: Cambridge U. Press.
Levitin, Daniel (2006). *This is Your Brain on Music*. NY: Dutton.
Martin, Rod (2007). *Psychology of Humor*. Boston: Elsevier Academic Press.
Mithen, Steven J. (2005). *The Singing Neanderthals*. London: Weidenfeld & Nicholson.
Murray, Charles (2003). *Human Accomplishment*. NY: HarperCollins.
Nettle, Daniel (2001). *Strong Imagination*. Oxford, UK: Oxford U. Press.
Pinker, Steven (1994). *The Language Instinct*. NY: Morrow.
Provine, Robert R. (2000). *Laughter*. NY: Viking.
Simonton, Dean K. (1999). *Origins of Genius*. NY: Oxford U. Press.
Strauss, Neil (2005). *The Game*. NY: ReganBooks.

## *Personality Traits*

Canli, Turhan (Ed.). (2006). *Biology of Personality and Individual Differences*. NY: Guilford.
Chamorro-Premuzic, Tomas, & Furnham, A. (2005). *Personality and Intellectual Competence*. Mahwah, NJ: Lawrence Erlbaum.
Funder, David C. (2006). *The Personality Puzzle*. NY: W. W. Norton.
Furnham, Adrian (2008). *Personality and Intelligence at Work*. Psychology Press.
Furnham, Adrian, & Heaven, P. (1999). *Personality and Social Behaviour*. London: Arnold.
John, Oliver P., et al. (2008). *Handbook of Personality*. NY: Guilford Press.
Matthews, Gerald, Deary, I., & Whiteman, M. (2004). *Personality Traits*. Cambridge, UK: Cambridge U. Press.
McCrae, Robert R., & Costa, P. T. (2003). *Personality in Adulthood*. NY: Guilford Press.
Millon, Theodore, et al. (Eds.). (2003). *Psychopathy*. NY: Guilford Press.
Nettle, Daniel (2007). *Personality*. Oxford, UK: Oxford U. Press.
Pervin, Lawrence A., & John, O. P. (Eds.). (1999). *Handbook of Personality Psychology*. NY: Guilford Press.

## Intelligence

Deary, Ian J. (2001). *Intelligence*. Oxford, UK: Oxford U. Press.

Deary, Ian J., et al. (2008). *A Lifetime of Intelligence*. Washington, DC: American Psychological Association.

Emery, Nathan, et al. (Eds.). (2008). *Social Intelligence*. NY: Oxford U. Press.

Geary, David C. (2005). *The Origin of Mind*. Washington, DC: American Psychological Association.

Herrnstein, Richard J., & Murray, C. (1994). *The Bell Curve*. NY: Free Press.

Jensen, Arthur R. (1998). *The g Factor*. London: Praeger.

Roberts, Maxwell J. (Ed.). (2007). *Integrating the Mind*. Hove, UK: Psychology Press.

Lynn, Richard (2008). *The Global Bell Curve*. Augusta, GA: Washington Summer Publishers.

Matthews, Gerald, et al. (2004). *Emotional Intelligence*. Cambridge, MA: MIT Press.

Sternberg, Robert J., & E. L. Grigorenko (Eds.) (2002). *The General Factor of Intelligence*. Mahwah, NJ: Lawrence Erlbaum.

Sternberg, Robert J., & Kaufman, J. C. (Eds.) (2002). *Evolution of Intelligence*. Mahwah, NJ: Lawrence Erlbaum.

## Behavior Genetics, Genomics, and Neurogenetics

Carey, Gregory (2002). *Human Genetics for the Social Sciences*. NY: Sage.

Jones, Byron C., & Mormede, P. (Eds.). (2006). *Neurobehavioral Genetics*. Boca Raton, FL: CRC Press.

Lesk, Arthur (2007). *Introduction to Genomics*. NY: Oxford U. Press.

Pagel, Mark, & Pomiankowski, A. (Eds.). (2007). *Evolutionary Genomics and Proteomics*. Sunderland, MA: Sinauer.

Parens, Erik, et al. (Eds.). (2005). *Wrestling with Behavioral Genetics*. Baltimore, MD: Johns Hopkins U. Press.

Plomin, Robert, et al. (2008). *Behavioral Genetics*. NY: Worth Publishers.

Ridley, Matt (2003). *Nature via Nurture*. NY: HarperCollins.

## Education, Credentialism, and Intelligence

Bok, Derek (2004). *Universities in the Marketplace*. Princeton, NJ: Princeton U. Press.

Carson, John (2007). *The Measure of Merit*. Princeton, NJ: Princeton U. Press.

Karabel, Jerome (2005). *The Chosen*. Boston: Houghton Mifflin.

Miyazaki, Ichisada (1981). *China's Examination Hell*. New Haven, CT: Yale U. Press.

Molnar, Alex (2007). *School Commercialism*. NY: Routledge.

Murray, Charles (2008). *Real Education*. NY: Crown Forum.

Phelps, Richard P. (Ed.). (in press). *The True Measure of Educational and Psychological Tests*. Washington, DC: American Psychological Association.

Springer, Sally P., & Franck, M. R. (2005). *Admission Matters*. San Francisco, CA: Jossey-Bass.

Veblen, Thorstein (1918). *The Higher Learning in America*. NY: B. W. Huebsch.

Wolf, Alison (2003). *Does Education Matter?* NY: Penguin Global.

## The Evolution of Morality, Cooperation, Trust, and Social Capital

Binmore, Ken (2005). *Natural Justice*. Oxford, UK: Oxford U. Press.

De Waal, Frans B. M. (2006). *Primates and Philosophers*. Princeton, NJ: Princeton U. Press.

Fukuyama, Francis (1996). *Trust*. NY: Free Press.

Gazzaniga, Michael S. (2005). *The Ethical Brain*. Washington, DC: Dana Press.

Hammerstein, Peter (Ed.). (2003). *Genetic and Cultural Evolution of Cooperation*. Cambridge, MA: MIT Press.

Hauser, Mark (2006). *Moral Minds*. NY: Ecco.

Lin, Nan (2001). *Social Capital*. NY: Cambridge U. Press.

McCullough, Michael E. (2008). *Beyond Revenge*. San Francisco, CA: Jossey-Bass.

Ostrom, Elinor, & Ahn, T. K. (Eds.). (2003). *Foundations of Social Capital*. Northampton, MA: Edward Elgar.

Putnam, Robert D. (2000). *Bowling Alone*. NY: Simon & Schuster.

Ridley, Matt (1996). *The Origins of Virtue*. NY: Penguin.

Sinnott-Armstrong, Walter (Ed.). (2008). *Moral Psychology*. Cambridge, MA: MIT Press.

Skyrms, Brian (2003). *The Stag Hunt and the Evolution of Social Structure*. Cambridge, UK: Cambridge U. Press.

Sober, Eliot, & Wilson, D. S. (1998). *Unto Others*. Cambridge, MA: Harvard U. Press.

Wilson, David S. (2003). *Darwin's Cathedral*. Chicago, IL: U. Chicago Press.


## Communities, Co-Living, New Urbanism, and Housing Issues

Chiras, Dan, & Wann, D. (2003). *Superbia*. Gabriola, BC, Canada: New Society Publishers.

Etzioni, Amitai (Ed.). (1998). *The Essential Communitarian Reader*. Landham, MD: Rowman & Littlefield.

Flint, Anthony (2006). *This Land*. Baltimore, MD: Johns Hopkins U. Press.

Girouard, Mark (1985). *Cities and People*. New Haven, CT: Yale U. Press.

Jacobs, Jane (1993). *The Death and Life of Great American Cities*. NY: Modern Library.

Low, Setha (2005). *Behind the Gates*. NY: Routledge.

Mulgan, Geoff (2008). *Living and Community*. London: Black Dog Press.

Mumford, Lewis (1961). *The City in History*. NY: Harcourt.

Nelson, Robert H. (2005). *Private Neighborhoods and the Transformation of Local Government*. Washington, DC: Urban Institute Press.

Schama, Simon (1996). *Landscape and Memory*. NY: Vintage.

Waldheim, Charles (Ed.). (2006). *The Landscape Urbanism Reader*. Princeton, NJ: Princeton Architectural Press.


## Conspicuous Charity, Ethical Investment, and Business Morals

Bornstein, David (2007). *How to Change the World*. NY: Oxford U. Press.

Friedman, Benjamin M. (2006). *The Moral Consequences of Economic Growth*. NY: Vintage.

Gini, Al, & Marcoux, A. M. (2008). *Case Studies in Business Ethics*. NY: Prentice-Hall.

Hancock, John (1999). *The Ethical Investor*. London: Financial Times/Prentice Hall.

Handy, Charles, & Handy, E. (2006). *The New Philanthropists*. London: William Heinemann.

Harrison, Rob, et al. (Eds.). (2005). *The Ethical Consumer*. Thousand Oaks, CA: Sage.

Hollender, Jeffrey, & Fenichell, S. (2003). *What Matters Most*. NY: Basic Books.

Tapscott, Don, & Ticoll, D. (2003). *The Naked Corporation*. NY: Free Press.

Vogel, David (2006). *The Market for Virtue*. Washington, DC: Brookings Institution Press.

Zak, Paul J. (Ed.). (2008). *Moral Markets*. Princeton, NJ: Princeton U. Press.

*Policy Implications, Consumption Taxes*

Crawford, Charles, & Salmon, C. (Eds.). (2004). *Evolutionary Psychology, Public Policy, and Personal Decisions*. Mahwah, NJ: Lawrence Erlbaum.

Doherty, Brian (2008). *Radicals for Capitalism*. NY: PublicAffairs.

Epstein, Richard A. (2003). *Skepticism and Freedom*. Chicago, IL: U. Chicago Press.

Frank, Robert H. (2007). *Falling Behind*. Berkeley: U. California Press.

Friedman, Milton (2002). *Capitalism and Freedom*. Chicago, IL: U. of Chicago Press.

Goodenough, Oliver R., & Zeki, S. (Eds.). (2006). *Law and the Brain*. NY: Oxford U. Press.

Kuran, Timur (1995). *Private Truths, Public Lies*. Cambridge, MA: Harvard U. Press.

McCaffery, Edward J. (2006). *Fair not Flat*. Chicago, IL: Chicago U. Press.

McCaffery, Edward J., & Slemrod, J. (Eds.). (2006). *Behavioral Public Finance*. NY: Russell Sage Foundation.

Mulgan, Geoff (2006). *Good and Bad Power*. London: Allen Lane.

Postrel, Virginia (1998). *The Future and its Enemies*. NY: Touchstone.

Rubin, Paul H. (2002). *Darwinian Politics*. New Brunswick, NJ: Rutgers U. Press.

Seidman, Laurence (1997). *The USA Tax*. Cambridge, MA: MIT Press.

Singer, Peter (2000). *A Darwinian Left*. New Haven, CT: Yale U. Press.

Somit, Albert, & Peterson, S. A. (Eds.). (2003). *Human Nature and Public Policy*. NY: Palgrave Macmillan.

Sowell, Thomas (2007). *A Conflict of Visions*. NY: Basic Books.

Tabb, William K. (2004). *Economic Governance in the Age of Globalization*. NY: Columbia U. Press.

Witt, Ulrich (2003). *The Evolving Economy*. Aldershot, UK: Edward Elgar.

Young, H. Peyton (1998). *Individual Strategy and Social Structure*. Princeton, NJ: Princeton U. Press.

## The Future of Consumerism

Gershenfeld, Neil (2005). *Fab*. NY: Basic Books.

Giddens, Anthony (2006). *Europe in the Global Age*. Malden, MA: Polity.

Gimore, James H., & Pine, B. J. (2000). *Markets of One*. Boston, MA: Harvard Business School Press.

Hutton, Will (2006). *The Writing on the Wall*. NY: Free Press.

Khanna, Tarun (2007). *Billions of Entrepreneurs*. Boston, MA: Harvard Business School Press.

Kurzweil, Ray (2005). *The Singularity is Near*. NY: Viking.

Meredith, Robyn (2008). *The Elephant and the Dragon*. NY: W. W. Norton.

Pine, B. Joseph, & Gilmore, J. (1999). *The Experience Economy*. Cambridge, MA: Harvard Business School Press.

Rifkin, Jeremy (2001). *The Age of Access*. NY: Tarcher/Putnam.

Rifkin, Jeremy (2004). *The European Dream*. NY: Tarcher/Putnam.

Sachs, Jeffrey (2008). *Common Wealth*. NY: Penguin.

Steffen, Alex (Ed.). (2006). *Worldchanging*. NY: Abrams.

Turner, Fred (2006). *From Counterculture to Cyberculture*. Chicago, IL: U. Chicago Press.

Von Hippel, Eric (2006). *Democratizing Innovation*. Cambridge, MA: MIT Press.

Wright, Robert (2001). *Nonzero*. NY: Vintage.

Young, Simon (2006). *Designer Evolution*. Amherst, MA: Prometheus Books.

Yunus, Muhammed (2007). *Creating a World Without Poverty*. NY: PublicAffairs.

## Previous Books by Geoffrey Miller

Miller, Geoffrey (2000). *The Mating Mind: How Sexual Choice Shaped the Evolution of Human Nature*. NY: Doubleday.

Geher, Glenn, & Miller, G. F. (Eds.). (2007). *Mating Intelligence: Sex, Relationships, and the Mind's Reproductive System*. Mahwah, NJ: Lawrence Erlbaum.

## Other Relevant Publications by Geoffrey Miller and Collaborators

Andrews, P. W., Gangestad, S. W., Miller, G. F., Haselton, M. G., Thornhill, R., & Neale, M. C. (2008). Sex differences in detecting sexual infidelity:

Results of a maximum likelihood method for analyzing the sensitivity of sex differences to underreporting. *Human Nature, 19,* 347–373.

Arden, R., Gottfredson, L., & Miller, G. F. (submitted). Does a fitness factor contribute to the association between intelligence and health outcomes? *Intelligence.*

Arden, R., Gottfredson, L., Miller, G. F., & Pierce. A. (in press). Intelligence and semen quality are positively correlated. *Intelligence.*

Geher, Glenn, Miller, G. F., & Murphy, J. (2007). Introduction: The origins and nature of mating intelligence. In G. Geher & G. Miller (Eds.), *Mating Intelligence: Sex, Relationships, and the Mind's Reproductive System* (pp. 3–34). Mahwah, NJ: Erlbaum.

Greengross, Gil, & Miller, G.F. (2008). Dissing oneself versus dissing rivals: Effects of status, personality, and sex on the short-term and long-term attractiveness of self-deprecating and other-deprecating humor. *Evolutionary Psychology, 6,* 393–408.

Griskevicius, V., Tybur, J. M., Sundie, J. M., Cialdini, R. B., Miller, G. F., & Kenrick, D. T. (2007). Blatant benevolence and conspicuous consumption: When romantic motives elicit costly displays. *Journal of Personality and Social Psychology, 93,* 85–102.

Haselton, Martie G., & Miller, G. F. (2006). Women's fertility across the cycle increases the short-term attractiveness of creative intelligence. *Human Nature, 17,* 50–73.

Hooper, Paul, & Miller, G. F. (2008). Mutual mate choice can drive ornament evolution even under perfect monogamy. *Adaptive Behavior, 16,* 53–70.

Kaufman, Scott, et al. (2007). The role of creativity and humor in mate selection. In G. Geher & G. Miller (Eds.), *Mating Intelligence: Sex, Relationships, and the Mind's Reproductive System* (pp. 227–262). Mahwah, NJ: Erlbaum.

Keller, Matthew C., & Miller, G. (2006). Which evolutionary genetic models best explain the persistence of common, harmful, heritable mental disorders? *Behavioral and Brain Sciences, 29,* 385–404.

Klimentidis, Y. C., Miller, G. F., & Shriver, M. D. (in press). Genetic admixture, self-reported ethnicity, self-estimated admixture, and skin pigmentation among Hispanics and Native Americans. *American Journal of Human Genetics.*

Mendenhall, Zack, & Miller, G. (in preparation). Conspicuous consumption in World of Warcraft: Auction versus vendor prices reveal the price premium for conspicuously cool weapons.

Miller, G. F. (1994). Beyond shared fate: Group-selected mechanisms for cooperation and competition in fuzzy, fluid vehicles. *Behavioral and Brain Sciences, 17,* 630–631.

Miller, G. F. (1997). Mate choice: From sexual cues to cognitive adaptations. In G. Cardew (Ed.), *Characterizing Human Psychological Adaptations (Ciba Foundation Symposium 208)* (pp. 71–87). NY: John Wiley.

Miller, G. F. (1997). Protean primates: The evolution of adaptive unpredictability in competition and courtship. In A. Whiten & R. W. Byrne (Eds.), *Machiavellian Intelligence II: Extensions and Evaluations* (pp. 312–340). Cambridge, UK: Cambridge U. Press.

Miller, G. F. (1998). How mate choice shaped human nature: A review of sexual selection and human evolution. In C. Crawford & D. Krebs (Eds.), *Handbook of Evolutionary Psychology: Ideas, Issues, and Applications* (pp. 87–129). Mahwah, NJ: Lawrence Erlbaum.

Miller, G. F. (1998). Waste is good. *Prospect* magazine, Feb., 18–23.

Miller, G. F. (1999). Sexual selection for cultural displays. In R. Dunbar, C. Knight, & C. Power (Eds.), *The Evolution of Culture* (pp. 71–91). Edinburgh, Scotland: Edinburgh U. Press.

Miller, G. F. (2000). Marketing. In J. Brockman (Ed.), *The greatest inventions of the last 2,000 years* (pp. 121–126). NY: Simon & Schuster.

Miller, G. F. (2000). Memetic evolution and human culture. *Quarterly Review of Biology, 75,* 434–436.

Miller, G. F. (2000). Mental traits as fitness indicators: Expanding evolutionary psychology's adaptationism. In D. LeCroy & P. Moller (Eds.), *Evolutionary Perspectives on Human Reproductive Behavior (Annals of the New York Academy of Sciences, Volume 907)* (pp. 62–74). NY: New York Academy of Sciences.

Miller, G. F. (2000). Sexual selection for indicators of intelligence. In G. Bock, J. Goode, & K. Webb (Eds.), *The Nature of Intelligence (Novartis Foundation Symposium 233)* (pp. 260–275). NY: John Wiley.

Miller, G. F. (2000). Evolution of human music through sexual selection. In N. L. Wallin, B. Merker, & S. Brown (Eds.), *The Origins of Music* (pp. 329–360). MIT Press.

Miller, G. F. (2000). Technological evolution as self-fulfilling prophecy.

In J. Ziman (Ed.), *Technological Innovation as an Evolutionary Process* (pp. 203–215). Cambridge, UK: Cambridge U. Press.

Miller, G. F. (2000). How to keep our meta-theories adaptive: Beyond Cosmides, Tooby, and Lakatos. *Psychological Inquiry, 11,* 42–46.

Miller, G. F. (2001). Aesthetic fitness: How sexual selection shaped artistic virtuosity as a fitness indicator and aesthetic preferences as mate choice criteria. *Bulletin of Psychology and the Arts, 2,* 20–25.

Miller, G. F. (2002). How did language evolve? In H. Swain (Ed.), *Big Questions in Science* (pp. 79–90). London: Jonathan Cape.

Miller, G. F. (2002). The science of subtlety. In J. Brockman (Ed.), *The Next Fifty Years* (pp. 85–92). NY: Vintage.

Miller, G. F. (2003). Fear of fitness indicators: How to deal with our ideological anxieties about the role of sexual selection in the origins of human culture. In *Being Human: Proceedings of a Conference Sponsored by the Royal Society of New Zealand (Miscellaneous series 63)* (pp. 65–79). Wellington, NZ: Royal Society of New Zealand.

Miller, G. F. (2006). The Asian future of evolutionary psychology. *Evolutionary Psychology, 4,* 107–119.

Miller, G. F. (2006). Asian creativity: A response to Satoshi Kanazawa. *Evolutionary Psychology, 4,* 129–137.

Miller, G. F. (2007). Sexual selection for moral virtues. *Quarterly Review of Biology, 82,* 97–125.

Miller, G. F. (2007). Brain evolution. In S. W. Gangestad & J. A. Simpson (Eds.), *The Evolution of Human Mind: Fundamental Questions and Controversies* (pp. 287–293). NY: Guilford Press.

Miller, G. F. (2007). Mating intelligence: Frequently asked questions. In G. Geher & Miller, G. F. (Eds.), *Mating Intelligence: Sex, Relationships, and the Mind's Reproductive System* (pp. 367–393). Mahwah, NJ: Erlbaum.

Miller, G. F. (2007). Reconciling evolutionary psychology and ecological psychology: How to perceive fitness affordances. *Acta Psycholigica Sinica, 39*(3), 546–555.

Miller, G. F. (2007). Runaway consumerism explains the Fermi paradox. In J. Brockman (Ed.), *What is Your Dangerous Idea?* (pp. 240–243). NY: Harper Perennial.

Miller, G. F. (2008). Magnaminity, fidelity, and other sexually-selected virtues. In W. Sinnott-Armstrong (Ed.), *Moral Psychology (Vol. 1): The Evo-*

*lution of Morality: Adaptations and Innateness* (pp. 209–243). Cambridge, MA: MIT Press.

Miller, G. F. & Penke, L. (2007). The evolution of human intelligence and the coefficient of additive genetic variance in human brain size. *Intelligence, 35,* 97–114.

Miller, G. F., & Tal, I. (2007). Schizotypy versus intelligence and openness as predictors of creativity. *Schizophrenia Research, 93,* 317–324.

Miller, G. F., & Todd, P. M. (1995). The role of mate choice in biocomputation: Sexual selection as a process of search, optimization, and diversification. In W. Banzhaf & F. H. Eeckman (Eds.), *Evolution and Biocomputation: Computational Models of Evolution* (pp. 169–204). Berlin: Springer-Verlag.

Miller, G. F., & Todd, P. M. (1998). Mate choice turns cognitive. *Trends in Cognitive Sciences, 2,* 190–198.

Miller, G. F., Tybur, J., & Jordan, B. (2007). Ovulatory cycle effects on tip earnings by lap-dancers: Economic evidence for human estrus? *Evolution and Human Behavior, 28,* 375–381.

Penke, Lars, Denissen, J. J. A., & Miller, G. (2007). The evolutionary genetics of personality. *European J. of Personality, 21,* 549–587.

Penke, L., Denissen, J. J., & Miller, G. F. (2007). Evolution, genes, and interdisciplinary personality research. *European Journal of Personality, 21,* 639–665.

Prokosch, Mark, Yeo, R., & Miller, G. (2005). Intelligence tests with higher g-loadings show higher correlations with body symmetry: Evidence for a general fitness factor mediated by developmental stability. *Intelligence, 33,* 203–213.

Sefcek, J. A., Brumbach, B. H., V·squez, G., & Miller, G. F. (2006). The evolutionary psychology of human mate choice: How ecology, genes, fertility, and fashion influence our mating behavior. *Journal of Psychology and Human Sexuality, 18,* 125–182.

Shaner, Andrew, Miller, G. F., & Mintz, J. (2004). Schizophrenia as one extreme of a sexually selected fitness indicator. *Schizophrenia Research, 70,* 101–109.

Shaner, Andrew, Miller, G. F., & Mintz, J. (2007). Age at onset of schizophrenia: Evidence of a latitudinal gradient. *Schizophrenia Research, 94,* 58–63.

Shaner, Andrew, Miller, G. F., & Mintz, J. (2007). Mental disorders as catastrophic failures of mating intelligence. In G. Geher & G. Miller (Eds.), *Mating Intelligence: Sex, Relationships, and the Mind's Reproductive System* (pp. 193–223). Mahwah, NJ: Erlbaum.

Shaner, Andrew, Miller, G. F., & Mintz, J. (2009). Autism as the low-fitness extreme of a parentally selected fitness indicator. *Human Nature, 19*, 389–413.

Todd, Peter, & Miller, G. F.(1999). From Pride and Prejudice to Persuasion: Satisficing in mate search. In G. Gigerenzer & P. Todd. (Eds.), *Simple Heuristics that Make us Smart* (pp. 286–308). Oxford, UK: Oxford U. Press.

Tybur, Josh, Miller, G. F., & Gangestad, S. (2007). Testing the controversy: An empirical examination of adaptationists' attitudes towards politics and science. *Human Nature, 18*, 313–328.

## Fiction

Novels and short story collections contain some of the best thinking and writing about consumerism and its alternatives. The authors below (followed by their most relevant works) seem the most consistently thought-provoking.

Martin Amis, *Success; Money; London Fields; The Information; Pornoland*

Margaret Atwood, *The Edible Woman; The Handmaid's Tale; Oryx and Crake*

J. G. Ballard, *Crash; High Rise; The Atrocity Exhibition; The Unlimited Dream Company; Cocaine Nights; Millenium People; Super-Cannes*

Nicholson Baker, *The Mezzanine; Room Temperature; Vox; The Fermata; A Box of Matches*

Iain M. Banks, *Consider Phlebas; The Player of Games; The State of the Art; Use of Weapons; Excession; Look to Windward; The Algebraist; Matter*

Douglas Coupland, *Generation X; Shampoo Planet; Microserfs; Girlfriend in a Coma; Life after God; Miss Wyoming; Jpod*

Philip K. Dick, *We can Build You; Confessions of a Crap Artist; A Scanner Darkly; Lies Inc.*

Bret Easton Ellis, *American Psycho; Glamorama*

William Gibson, *Neuromancer; Virtual Light; Pattern Recognition; Spook Country*

Aldous Huxley, *Brave New World; Island*

Jeff Noon, *Vurt; Pollen; Nymphomation; Pixel Juice*

Chuck Palahniuk, *Fight Club; Survivor; Choke; Lullaby; Diary; Haunted; Rant; Snuff*

Frederick Pohl, *Midas World*

Neal Stephenson, *Interface; Snow Crash; The Diamond Age, Anathem*

Irvine Welsh, *Trainspotting; Ecstasy; The Acid House; Porno; Filth*

## Films

Many films have provocatively and insightfully addressed various issues in the psychology of consumerism, marketing, product design, self-display, status signals, and prehistoric versus modern life, such as:

*Amélie* (2001), *American Beauty* (1999), *American Psycho* (2000), *Apocalypto* (2006), *Artificial Intelligence* (2001), *A Scanner Darkly* (2006)

*Being John Malkovich* (1999), *Best in Show* (2000), *Blade Runner* (1982), *Boiler Room* (2000), *Boogie Nights* (1997), *Bowling for Columbine* (2002), *Brazil* (1985), *Breast Men* (1997)

*Cast Away* (2000), *Catch Me If You Can* (2002), *Charlie and the Chocolate Factory* (2005), *The Corporation* (2004)

*The Devil Wears Prada* (2006)

*Equilibrium* (2002), *Eternal Sunshine of the Spotless Mind* (2004), *Existenz* (1999)

*Fight Club* (1999)

*The Game* (1997), *Gattaca* (1997), *Glengarry Glen Ross* (1992), *Groundhog Day* (1993)

*I, Robot* (2004), *Idiocracy* (2006), *The Insider* (1999), *Into the Wild* (2007), *The Island* (2005)

*Jerry Maguire* (1996)

*Koyaanisqatsi* (1982)

*Little Miss Sunshine* (2006), *Lord of War* (2005), *Lost in Translation* (2003)

*Magnolia* (1999), *Manufacturing Consent* (1992), *The Matrix* (1999), *Mean Girls* (2004)

*Requiem for a Dream* (2000), *Robocop* (1987)

*Scarface* (1983), *The Shape of Things* (2003), *Sideways* (2004), *Strange Days* (1995)

*The Tao of Steve* (2000), *Thirteen* (2002), *The Truman Show* (1998), *Total Recall* (1990), *Trainspotting* (1996)

*Wall Street* (1987), *WALL-E* (2008), *What Women Want* (2000)

*The Yes Men* (2003)

# Acknowledgments

Thanks for everything to my editors, Rick Kot, Drummond Moir, Alban Miles, and Ravi Mirchandani, and to my agents, Katinka Matson and John Brockman.

For reading the entire book in draft form, and giving very useful feedback, very special thanks to: Rosalind Arden, Henry Baker, Jaime Confer, Chris Eppig, Carolyn Miller, Frank Miller, Daniel Nettle, Lars Penke, Carin Perilloux, Catherine Salmon, Andrew Shaner, Jill Sundie, Joshua Tybur, and Michael Wewerka.

For giving helpful feedback on particular sections of the book, thanks to: Nelson Amaral, Atalanta Arden-Miller, John Baker, Richard Baker, Michael Church, Dylan Evans, Vladas Griskevicius, Ori Heffetz, Mia Kersting, Emily Maple, Leslie Merriman, Gad Saad, Randy Thornhill, and Rhiannon West.

Thanks to the many other friends, colleagues, researchers, relatives, and students who have discussed or corresponded about this book's ideas with me over the past decade, including: Paul Andrews, Tim Bates, Susan Blackmore, Jim Boone, David Buss, David Byrne, Ann Caldwell, Richard Conniff, Helena Cronin, Oliver Curry, David DeAngelo, Ian Deary, Damon DeLaszlo, Denis Dutton, Vince Egan, Rachael Falcon, Robert Frank, David Funder, Steve Gangestad, John Gardner, Glenn Geher, Gerd Gigerenzer, Dan Goldstein, Oliver Goodenough, Linda Gottfredson, David Hall, Richard Harper, Martie Haselton, Paul Hooper, Nicholas Humphrey, Chris Jenkins, Satoshi Kanazawa, Hilly Kaplan, Anat Keinan, Matt Keller, Douglas Kenrick, Rebecca LeBredonchel, Andrea Levine, Gary McGovney, Zack Mendenhall, Ravi Mirchandani, Randy Nesse, Michael Norton, David O'Hanlon, John Orbell, Nando Pelusi, Steven Pinker, Matt Ridley, Andrew Shaner, Aubrey Sheiham, Catherine Randall, Peter Singer, Walter Sinnott-Armstrong, Ilanit Tal, Andy Thompson, Peter Todd, Paul Watson, Richard Webb, Ian Wedde, and Ron Yeo.

# Index